N/C MACHINABILITY
DATA SYSTEMS

N/C MACHINABILITY DATA SYSTEMS

Noel R. Parsons Editor

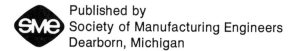

Published by
Society of Manufacturing Engineers
Dearborn, Michigan

Numerical Control Series

N/C Machinability Data Systems

Copyright © 1971 by the Society of Manufacturing Engineers, Dearborn, Michigan

Library of Congress Catalog Card Number: 74-153852

International Standard Book Number: 0-87263-029-3

First Edition   **Manufactured in the United States of America**

PREFACE

Numerically controlled machine tools must be used efficiently to provide the economy for which they were designed. An important factor in using such tools effectively is the efficient collection and use of accurate, reliable machinability data. Until now, the method of collecting machinability data, and the quality of the data used, has often been left up to individuals within each company. In many cases, there has been no coordination of information between those individuals, nor has there been a standard format for collecting machinability data within a company. Other companies rely primarily upon handbook figures as starting points for a machining operation and adapt those figures to conditions as work progresses.

These informal methods of collecting and using machinability data are inefficient with comparatively inexpensive conventional machines, and for costly N/C machines they are completely out of the question. For N/C machines, machinability data must be readily available and must be accurate. Down time caused by slow machinability data retrieval and application can turn a profitable N/C operation into a losing one. Accidents to machines, tools, or parts because of incorrect data are extremely expensive as well.

Today, many companies are working towards the optimization of their N/C operations by collecting and using machinability data systematically—using computers to store and recall the data and even to give cost analyses of their operations. The day of the experienced machinist who adjusts machining conditions to solve problems as they occur will soon be gone. Machinability data systems are becoming more and more a science and less and less arbitrary experimentation.

The Society is presenting this book for three reasons: (1) to demonstrate the importance of having machinability data in an organized form (not only because N/C demands it, but also because conventional machining operations can benefit from it); (2) to present the practical approaches currently being employed to collect and use machinability data; and (3) to present the advanced concepts of collecting and using machinability data, including the computerized systems.

The authors have tried to present as much detail as possible about the machinability data systems now being used or developed. However, as you read and study the concepts presented here it will become evident that very little information is available because very little work has been done in this field. The Society hopes that this book, by presenting available, current information on machinability data systems in one source, will stimulate ideas for improving existing machinability data systems and will promote research into new and better concepts. In today's highly competitive manufacturing community, the availability of machinability data may determine the success or failure of companies involved in machining either by conventional or numerically controlled methods.

SME wishes to extend its appreciation to S. B. L. Wilson, Manager, N/C Computer Group, and John Taylor, Manager, Advanced Aerofoil Facility, of Rolls-Royce Limited, and to John Maranchik of John Maranchik and Associates for their aid in the preparation of the manuscript of this book. The Society also appreciates the help provided by William W. Gilbert, Manager, Machining Development Operations, General Electric Company; Karl W. Meyer, IBM Corporation; and Peter R. Artz, Senior Manufacturing Engineer, McDonnell Douglas Aircraft Corporation, in reviewing the manuscript and contributing toward its technical accuracy.

NOEL R. PARSONS

Dearborn, Michigan
January, 1971

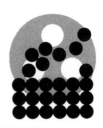

CONTENTS

CHAPTER 1   MACHINABILITY DATA FOR OPTIMIZED
N/C PRODUCTION   1

Machinability Data for N/C vs. Conventional
   Machining   2
Specialized Data for N/C   3
Machinability Data Interrelationships   10

CHAPTER 2   TYPICAL DATA SYSTEMS FOR N/C
PROGRAMMING   15

Data and Data Formats   15
Machinability Data Acquisition   23
Format and Program Preparation   35
Stringency of Programs   40
Data Record Keeping   60
Optimizing N/C Data and Formats   62

CHAPTER 3   COMPUTERIZED MACHINABILITY DATA
SYSTEMS   63

General Electric Data System   64
FAST for Feed and Speed Technology   85
EXAPT System   96
Abex System   114
Establishing a Computerized System   125

CHAPTER 4   COMPUTER ANALYSIS OF COST AND
PRODUCTION RATES   129

Cost and Production Analysis Methods   129
Machinability Data Storage and Retrieval   130
Computation of Cost and Operation Times   140

Cost and Production Analysis 144
Determining Optimum Cutting Conditions 152
Developing a Total System 154

CHAPTER 5   FUTURE MACHINABILITY DATA CONCEPTS   157

Role of the Computer 157
Optimizing the Cutting Process 160
New Uses of Machinability Data 166

CHAPTER 6   MACHINABILITY DATA: AVAILABILITY
AND ECONOMICS   169

Material-Removal Economy 169
The Machinability Index 172
Machinability Tests 173
Tool-Life Tests 177
Machinability Data for Numerical Control 184

APPENDIX   AIR FORCE MACHINABILITY DATA CENTER   189

BIBLIOGRAPHY   191

INDEX   197

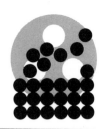

FIGURES

Figure 1-1   N/C machine-tool utilization data log.   4

Figure 1-2   Typical N/C operation sequence flow chart.   6

Figure 1-3   Two-axis, single-spindle N/C machine.   7

Figure 1-4   Two-axis N/C contouring lathe.   8

Figure 1-5   Three-axis, three-spindle "bridge-type" N/C contouring machine.   8

Figure 1-6   Adjustable cutting-tool holders.   9

Figure 1-7   Horizontal boring, drilling, and milling machine.   11

Figure 1-8   Precision positioning and contouring machining center.   12

Figure 1-9   Tap life vs. tap design for A-286 stainless steel.   13

Figure 1-10   Effect of side cutting-edge angle on tool life for turning.   13

Figure 2-1   ADAPT/AUTOSPOT software system flow chart.   16

Figure 2-2   Typical format and data for a two-axis N/C machine.   17

Figure 2-3   Portion of a descriptive operation and routing sheet for a three-axis N/C contouring profiler.   19 ·

Figure 2-4   Drawing of the part corresponding to the descriptive routing of Fig. 2–3.   20

Figure 2-5   Format and data of a slope and arc program for a three-axis-positioning N/C machining center with auxiliary equipment.   22

Figure 2-6   Nomographs for surface speed, torque, tangential load, horsepower, and metal-removal rate for milling.   24

Figure 2-7   Effect of microstructure in carbon steels as related to machinability.   37

Figure 2–8   Machinability of cast iron.   45

Figure 2–9   Examples of machinability data and ratings for
                  various steels.   54

Figure 2–10   Example of machinability data taken from the
                  Machining Data Handbook.   55

Figure 2–11   Sample part and cutter path drawings for a part
                  to be programmed using ADAPT.   58

Figure 2–12   ADAPT part program manuscript for the part shown in
                  Fig. 2–11.   59

Figure 2–13   Part program output listing for the part shown in
                  Fig. 2–11.   60

Figure 2–14   Program output listing of all calculations required to
                  control the cutter path shown in Fig. 2–11.   61

Figure 3–1   Tool life vs. cutting speed showing typical
                  tool-life lines.   66

Figure 3–2   Log-log plot correlating cutting speed and Brinell
                  hardness numbers for steels.   66

Figure 3–3   Log-log plot correlating cutting speed and Brinell
                  hardness numbers for cast iron.   66

Figure 3–4   Machinability ratings vs. Brinell hardness numbers.   67

Figure 3–5   Speed curves for B1112 steel and another material.   67

Figure 3–6   Results of studies of the effects of cutting fluids
                  on the cutting-speed/tool-life relationship.   68

Figure 3–7   Cutting speed vs. tool life as related to workpiece
                  surface finish.   68

Figure 3–8   Flank wear as a function of time for various
                  cutting speeds.   69

Figure 3–9   Log-log plot of speed vs. flank wear for
                  constant time.   69

Figure 3–10   Effect of feed on cutting speed.   69

Figure 3–11   Relationship between nose radius, surface finish,
                  and feed.   74

Figure 3–12   Relationship between depth of cut and cutting speed.   74

Figure 3–13   Profile factor as a function of nose radius,
                  side cutting-edge angle, and depth of cut.   75

Figure 3–14   Costs per piece vs. cutting speed.   75

Figure 3–15   Hi-E range curves.   78

Figure 3–16   Relationship between unit horsepower and
                  Brinell hardness numbers.   79

Figure 3–17   Input data sheet for use with the General Electric
               computerized machinability data system.   80

Figure 3–18   Detail of input data sheet—workpiece and
               cutting-tool data section.   81

Figure 3–19   Detail of input data sheet—machine-tool data section.   81

Figure 3–20   Computer printout of information when cemented
               carbide tool grade is dictated.   82

Figure 3–21   Detail of input data sheet—Hi-E Data for
               Disposable Inserts section.   83

Figure 3–22   Detail of input data sheet—Hi-E Data for
               Brazed Tools section.   83

Figure 3–23   General Electric data system computer printout
               of information resulting from tool-life insertion.   84

Figure 3–24   Computer printout of information resulting from
               request for Hi-E data.   85

Figure 3–25   Data outputs of the FAST computer-aided data systems.   86

Figure 3–26   FAST general system flow chart.   87

Figure 3–27   FAST material file listing.   88

Figure 3–28   Interpolation of the feed vs. drill diameter curve to
               determine drilling feed.   88

Figure 3–29   FAST machine reference file listing.   89

Figure 3–30   Detailed FAST machine file listing.   89

Figure 3–31   FAST tool file listing in numerical order by
               code number.   90

Figure 3–32   FASTER tool geometry file listing.   91

Figure 3–33   FAST tool file listing by tool diameter, tool type, and
               operation.   92

Figure 3–34   Printout of typical input data for the FAST system.   93

Figure 3–35   FAST initial interrogation output document.   93

Figure 3–36   FAST initial interrogation remarks output document.   94

Figure 3–37   FAST recommended feeds and speeds output document.   94

Figure 3–38   Tool interrogation input data printout for the FAST system.   95

Figure 3–39   FAST recommended feeds and speeds for the tools
               interrogated in Fig. 3–38.   95

Figure 3–40   FASTEST interrogation and output data printout.   96

Figure 3–41   FASTEST output data including estimated
               operation times.   96

Figure 3–42   Steps in industrial computer-aided manufacturing
                   with EXAPT.   99

Figure 3–43   General EXAPT 2 processor flow chart.   100

Figure 3–44   EXAPT materials file index card.   101

Figure 3–45   EXAPT cutting-tool file index card with tool
                   data inserted.   102

Figure 3–46   Calculation of feeds and speeds for EXAPT 1.   102

Figure 3–47   Turning-tool dimensions as entered in the
                   EXAPT tool file.   103

Figure 3–48   Engineering drawing of a lathe-machined shaft.   106

Figure 3–49   EXAPT 2 part program for the example shaft.   106

Figure 3–50   Sequence of operations and cuts for the example shaft. 107

Figure 3–51   Flow chart showing calculation of depth of cut, feeds,
                   and speeds in EXAPT 2.   110

Figure 3–52   Graph showing determination of feeds in EXAPT 2.   112

Figure 3–53   Graph showing determination of speeds in EXAPT 2.   112

Figure 3–54   Technological evaluation program for creating and
                   testing material data files for EXAPT.   112

Figure 3–55   Printout of automatic calculation of cutting
                   values in EXAPT 2.   113

Figure 3–56   EXAPT 2 computer graph printout of automatically
                   calculated cutting values.   114

Figure 3–57   General Abex data system flow chart.   117

Figure 3–58   Detailed Abex machinability analysis for a
                   turning operation.   119

Figure 3–59   Detailed Abex machinability analysis for a
                   drilling operation.   120

Figure 3–60   Detailed Abex machinability analysis for an
                   end-milling operation.   120

Figure 3–61   Abbreviated flow chart showing Abex SEARCH
                   and COMPUTATION blocks.   123

Figure 3–62   Sample FORTRAN programming of the flow chart
                   of Fig. 3–61.   124

Figure 3–63   "Milepost" project planning schedule for establishment
                   of a computerized machinability data system.   126

Figure 4–1   Format for the collection of tool-life data for turning. 131

Figure 4–2   Format for the collection of tool-life data for
                   face milling, end-mill slotting, and
                   peripheral end milling.   132

Figure 4–3   Format for the collection of tool-life data for
               drilling, reaming, and tapping.   133

Figure 4–4   Sample computerized file listing of N/C
               machining data for turning.   134

Figure 4–5   Sample computerized file listing of N/C
               machining data for drilling.   135

Figure 4–6   Lathe tools and setup.   138

Figure 4–7   Milling cutters and setup.   138

Figure 4–8   Drills, reamers, and their setup.   138

Figure 4–9   Taps and their setup.   139

Figure 4–10   Computer printout of N/C machining costs and
               operation times for a shaft.   143

Figure 4–11   Shaft to be machined on an N/C lathe.   144

Figure 4–12   Time-study and cost data for the shaft of Fig.
               4–11.   146

Figure 4–13   Tool-life data for turning, drilling, and tapping a part
               on an N/C lathe.   147

Figure 4–14   Part to be machined on an N/C machining center.   148

Figure 4–15   Time-study and cost data for the part shown in Fig. 4–14. 149

Figure 4–16   Tool-life data for face milling, peripheral end milling,
               end-mill slotting, and drilling on an N/C
               machining center.   150

Figure 4–17   Computer printout of N/C machining costs and operation
               times for the part shown in Fig. 4–14.   151

Figure 5–1   Machining costs as functions of cost per piece vs.
               cutting speed or production rate.   158

Figure 5–2   Flow chart for a drilling operation with a computerized
               machinability data system.   159

Figure 5–3   Simplified flow chart of a conventional N/C
               operation sequence.   161

Figure 5–4   Flow chart of direct numerical control (DNC) operations. 162

Figure 5–5   Relationship of an adaptive control to an N/C control
               unit and machine tool.   163

Figure 5–6   Optimization of production by reduction of the effects
               of common process variables with adaptive control.   164

Figure 5–7   Simple feedback type of adaptive control system.   164

Figure 5–8   Adaptive constraint control system.   165

Figure 5–9   Programmed adaptive control system.   166

Figure 5–10   Optimal adaptive control system.   166

Figure 5–11   Comparative machining costs for N/C and conventional
                   machining as functions of cost per piece vs. cutting
                   speed or production rate.   167

Figure 6–1   Growth in carbide tool shipments from 1961 through 1969.   172

Figure 6–2   Force measurement methods for the mechanical lathe
                   dynamometer.   174

Figure 6–3   Force measurement methods for a drill dynamometer.   175

Figure 6–4   Tool-life curves for face milling cast iron with a carbide
                   cutter.   176

Figure 6–5   Tool-life curves for turning AISI 4340 steel with
                   carbide tools.   177

Figure 6–6   Tool-life curves for turning thermal-resistant
                   Udimet 500 at 360 BHN using three cutting fluids
                   with an HSS tool.   178

Figure 6–7   Effect of feed on tool life in face milling a thermal-
                   resistant alloy at 42 R_c.   179

Figure 6–8   Results of tapping titanium alloy at 400 BHN
                   with three taps.   179

Figure 6–9   Wear land measurement in tool-life testing.   180

Figure 6–10   Example of profile end-milling data from the
                   Machining Data Handbook.   181

Figure 6–11   Example of drilling data from the
                   Machining Data Handbook.   185

Figure 6–12   Machining costs for hard-to-machine materials.   187

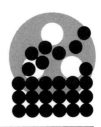

TABLES

Table II–1	Average Unit Power Requirements.	27
Table II–2a	Classification of Materials (Steels).	28
Table II–2b	Classification of Materials (Cast Iron).	29
Table II–2c	Classification of Materials (Copper Alloys, Brass, Bronze).	30
Table II–3	Speeds, Feeds, and Times for Drilling, Reaming, Counterboring, and Core Drilling.	32
Table II–4	Speeds and Feed Times for Tapping.	36
Table II–5	Speeds and Feeds for Side Milling and Plain Milling.	38
Table II–6	Speeds and Feeds for End-Mill Slotting—Machinability Classification 4B.	40
Table II–7	Speeds and Feeds for End Milling Using Side of Cutter —Machinability Classification 4B.	42
Table II–8	Speeds and Feeds for Face Milling with T.C. Full Back Cutters.	44
Table II–9	Speeds and Feeds for Open Boring on Machinability Classification 4B.	46
Table II–10	IBM Division Standard—Machinability Classifications.	48
Table II–11	IBM Division Standard—Speeds and Feeds for Plain Drilling.	51
Table II–12	IBM Division Standard—Speeds and Feeds for Milling.	56
Table III–1	Starting Cutting Speeds and Insert Grade Recommendations for Approximately 30 Min Tool Life.	70
Table III–2	Typical Machinability Ratings.	73
Table III–3	Effects of Tolerance on Allowable Flank Wear.	82
Table III–4	Features of the EXAPT Programming System.	98

Table IV–1   Symbols for Cost and Production Rate Equations for
              N/C Machining.   136

Table IV–2   Generalized Cost Equations for N/C Machining.   140

Table IV–3   Generalized Equations for Operation Time Per Piece
              and Production Rate for N/C Machine Tools.   142

Table IV–4   Optimized Cutting Speed Equations for Turning,
              Milling, Drilling, Reaming, and Tapping.   153

Table IV–5   Optimized Tool Life Equations for Turning, Milling,
              Drilling, Reaming, and Tapping.   153

Table VI–1   Estimate of United States Metal-Cutting
              Operations, 1958.   170

Table VI–2   Metal-Cutting Statistics for the United States, 1968.   170

Table VI–3   Values of Shipments of Metal-Cutting Tools by Class of
              Cutting Tool and Tool Material, 1968.   171

Table VI–4   Wear Land Widths for Testing Various Operations.   180

Table VI–5   Recommended Cutting Conditions for Machining René 41
              Solution-Treated to 321 BHN.   182

Table VI–6   Survey of Hard-to-Machine Material Usage
              Percentages, 1964–74.   187

MACHINABILITY DATA FOR OPTIMIZED N/C PRODUCTION

George J. Hoffmann, Ford Motor Company

A good machining process, like any sound business operation, is based on a foundation of planning, coordination, and feedback. The successful installation of numerical control, and its application to the machining process, requires the same foundation. The elements of sound engineering management are even more critical for N/C techniques than for conventional, nonautomated machining methods. The high cost and productive capacity of N/C equipment require strict discipline and efficient organization in order to achieve an attractive return on the original investment.

Machinability, or the "ease" with which a material may be worked, must be judged by several criteria. Any information which contributes to the control of these criteria may be considered as a variable that affects the engagement of tool and workpiece. The four most frequently used and universally understood criteria are the following:

1) Tool Life—If tool life on any operation is short, machine down time for tool change will be frequent, thereby affecting the production quantity per unit time and the cost per piece. If tool life is too long, then the cutting speed may be too slow, meaning that the number of pieces machined per unit time is low and that the cost per piece remains high.

2) Surface Finish—If the surface finish, even on a semifinished cut, is not controlled, then tools used on subsequent operations must machine severely deformed, work-hardened material. Tool life may then be decidedly reduced.

3) Accuracy—Cutting tools and cutting conditions must be selected which will satisfy the accuracy requirements of the part drawing.

4) Power Consumption—The volume of material to be removed, when equated to meaningful horsepower values, is necessary data for the selection of adequate machine-tool horsepower. The results of inadequate horsepower may be a stalled machine, broken tools, and the destruction of valuable workpieces.

The development of systems for machinability data and formats is a vital part of engineering for N/C. Such data systems provide the means of (1) establishing consistent operation of N/C machine tools, (2) ensuring an adequate supply of machining data to allow the maximum use of N/C machines, and (3) producing quality products

1

at a minimum cost. A machinability data system may be a manual reference type or it may be computerized. Whichever type it is, it should contain at least the following data:

1) Machinability data formats for N/C parts programming
2) Machining data including feeds, speeds, and tool-life parameters
3) A library of tooling data (tool geometry and predetermined set lengths)
4) Methods of holding tools.

MACHINABILITY DATA FOR N/C VS. CONVENTIONAL MACHINING

There is no major difference between the kind of machinability data used for N/C processing and the kind used for conventional manufacturing. The objectives of each method of manufacturing are the same. Both strive for size, shape, accuracy, surface integrity, and efficient production rates. The machinability data to achieve these objectives stem from the same sources—research and experience.

There is no difference in the deformation and shear phenomena by which material is removed in both types of manufacturing. The value of accurate, reliable information about tool material applications, optimum tool geometries, machine tool settings, and material variables is as important to conventional methods as to N/C methods.

On the other hand, N/C processing techniques often vary from conventional methods, for with N/C machines tool life may be increased, damage to tools and workpieces may be reduced, and piece-to-piece time can usually be shortened. These advantages are the result of complete, accurate motion control in which manual machining methods are largely eliminated. It is this control that makes N/C worthwhile and for which most N/C users have found justification. It is also this control that generally makes simple data systems inadequate for N/C use.

Simple Data Systems

The degree of sophistication of any machinability data system depends on the user organization's facilities and size. But any organization that is planning for N/C should recognize the true value of simple systems as starting points upon which firm foundations of machinibility data may be constructed. The greatest danger of simple systems lies in the fact that their information often becomes too generalized.

Many individuals and organizations have contributed to the development of machining data systems, but their data, as published for reference, are determined from unique examples that are then sometimes extrapolated to cover a wide range of applications. In order to publish data in formats that are readily accessible, readable, and understandable, detailed information must be kept to a minimum. Thus, the unique parameters upon which the data are based are often lost to the ultimate user. A manufacturing engineer, through his personal experience and machining knowledge, can use his own preferred source of generalized information in his daily work with a high degree of confidence.

This ability to work with generalized machinability data is not always efficient for the operation of N/C machine tools, however. Simple machining data systems may not provide the information needed to prepare correct N/C part programs without excessive machine inactivity. Because of the high cost of N/C machining time—as high as fifteen times that of conventional machining time in some cases—down time on N/C machine tools cannot be borne without undue sacrifice of profits. Use of generalized machining data may cause production delays while it is analyzed for accurate parameters to be used in a particular program. If generalized data is used in a program, delays may also result when the program is tried on the machine and corrections must be made.

SPECIALIZED DATA FOR N/C

There are five main reasons why simple machinability data systems are usually inadequate for N/C production and why specialized data systems should be developed for N/C. These reasons are the following:

1) The higher cost of N/C machine tools and machining time
2) Programming costs
3) Variations in N/C machine tools and N/C production requirements
4) The special relationships of the workpiece to the N/C machine
5) The shift of responsibility for N/C operations from operator to programmer.

N/C Cost vs. Utilization

Anyone who has purchased or investigated the purchase of an N/C machine tool will know that the costs of these machines, depending on the type, usually range from two to as much as ten times more than the costs of most equivalent conventional machine tools. Because of these high costs, N/C machines must be used efficiently. N/C utilization must be continually compared to the original investment and must be closely controlled and held at high levels to justify that investment. The following are two specific methods of obtaining the most efficient N/C utilization:

1) Development of specialized data systems
2) Development of methods of work content study.

Specialized Data Systems. Later chapters will discuss specialized data systems for N/C machining in greater detail, but at this point it is important to recognize their value in controlling N/C utilization. Specialized formats should be developed for the following data needs:

1) Parts programming
2) Feeds and speeds
3) Cutting-tool libraries
4) Toolholders and adaptors
5) Machine utilization reporting.

The last item in this group, a specialized data format for reporting N/C machine utilization, is most important for the development of maximum machine-tool use and for the justification of capital investment for additional N/C equipment. This data format should present a running account of all parts, tools, codes for machine activity and inactivity, starting and stopping times, and so on, to show a complete part and machining history. These data may then be collected, possibly organized by the company's data system, and analyzed to point out phases of machining time which can be made more productive or areas of support which may be improved (1, 2, 3, 4). A typical reporting form for N/C machine-tool utilization is shown in Fig. 1-1.

Work Content Study Methods. After a part has been designed, engineering preparation for its production begins with the analysis of work content to select the proper work for N/C machining. When the elements of part work content are compared to N/C machine-tool capabilities and associated operation costs, the proper equipment

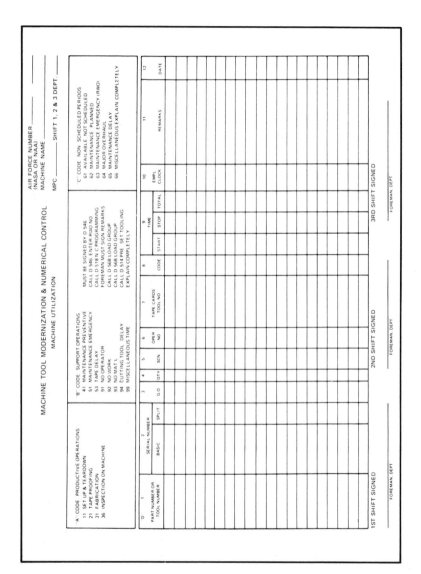

Fig. 1-1.—N/C machine-tool utilization data log (1).

can be selected to ensure full use of machine tools for maximum production at the lowest possible cost. Methods of analyzing work content should be developed. Engineering drawings should be studied in detail, and the machining requirements of the part should be compared to machining elements. Some work content elements that should be considered for each part are as follows:

1) Number of operations to be performed
2) Accuracy required
3) Surface finishes required
4) Number of tools to be used
5) Required part rotation
6) Material to be used
7) Any engineering changes expected
8) Special features required.

Parts Programming Cost

The cost of developing a part program itself is another reason why simple data systems are often inadequate for N/C use. An N/C tape, or whatever input medium is used, is the result of time-consuming thought and work as the efforts of the product designer are translated from lines on an engineering drawing to coded information on the final tool control medium. The entire process from drawing to control tape, as shown in Fig. 1–2, is a predetermined sequence, and the high cost of engineering that sequence demands complete and correct supporting data.

Machine-Tool and Production Variations

The wide variety of N/C machine tools available also makes the use of more sophisticated data systems desirable. A common source of machinability data can be applied to different parts on different N/C machine tools just as it can to conventional machining operations. Feeds and speeds used are influenced most by workpiece material and cutting tools and less by the type of machine used. However, the manufacturing engineer must also be constantly aware of the variations in N/C machine-tool configurations and capabilities. Drilling a simple hole with a two-axis N/C drill (Fig. 1–3), for example, is an entirely different problem from using a multiaxis machining center (Fig. 1–5) for the same job. Use of N/C means that each particular machining problem requires a different method of handling machining data.

When the two-axis drill is used, cutting tools need not be compromised; short, stubby tools can be used with maximum speeds and feeds. On the other hand, the machining center, which is designed for complex work, may require the use of longer, less rigid cutting tools. The machining center can perform the same simple drilling work, but it must do it with more complex operations in order to reduce tooling costs, handling time, and floor space. The manufacturing engineer must be alert to recognize where compromises in machining data should be made to reduce costs.

In the same way, N/C production requirements may prohibit the use of simple machinability data systems. As an example, a shop dealing mainly in one-of-a-kind or short-run jobs may elect to standardize certain types of cutting tools to reduce costs. A shop engaged in long production runs, however, would not be so concerned with reducing cutting-tool costs and would require machinability data applicable to each type of cutting tool used. The requirements of each type of production should be analyzed to determine whether and where compromises in the use of machinability data should be

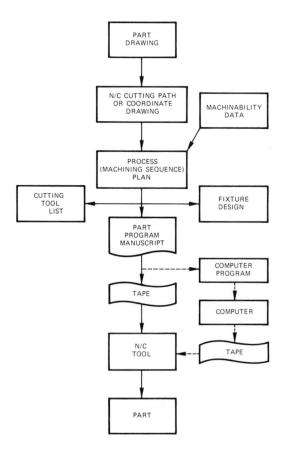

Fig. 1-2.—Typical N/C operation sequence flow chart.

made. If a compromise should be reached at the expense of tooling cost or machine cycle time, the manufacturing engineer should learn to recognize where the greater benefits lie. There is no set rule for such decisions, and the answer will be different for each application.

Workpiece–Machine Relationships

The relationships of the workpiece to the machine also help to establish the need for specialized machinability data for N/C. These relationships are:

1) Part configuration and machine configuration
2) Rigidity of parts, fixtures, and tools
3) Part material and tool life
4) Part producibility and machine configuration.

Part Configuration. Configuration of the part is of primary importance to machinability data, for the size, shape, and work content in deriving that shape are the parameters upon which machine-tool selection is based. Some parts require an N/C

Fig. 1-3.—Two-axis, single-spindle N/C machine. *(Courtesy, Cincinnati Milacron, Inc.)*

machine dedicated only to one machining operation such as drilling, milling, or turning. Other parts require a machining center with several axes, continuous path capability, and automatic tool changing. While part configuration alone may dictate one type of machine, the overall operating objectives of the company may show that a given part should pass over a number of different types of N/C machines in order to maximize productivity or improve quality. Specialized data systems will be able to provide the information necessary to select the right N/C machines and machining operations.

Rigidity. The rigidity factors of the part, fixturing, and cutting tools may change from machine to machine and from setup to setup, and each combination of these factors will affect the use of machinability data. For example, although aluminum parts may be machined at much higher feeds and speeds than similar steel parts, the modulus of elasticity of aluminum is only one-third that of steel, and the forces resulting from this lower elasticity may cause cutting-tool deflection problems. Such problems must be predicted by the machinability data available so that corrective programming can be planned (5). Machinability data should also be specialized to make allowances for differences in rigidity between different fixture setups or even different machine tools.

Tool Life. Tool life, a factor which can easily be changed by N/C part programming, would be considered differently in a prototype shop than in a production shop.

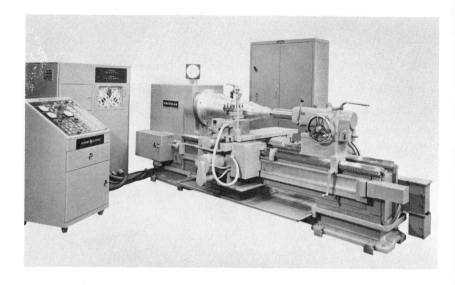

Fig. 1–4.—Two-axis N/C contouring lathe. *(Courtesy, The American Tool Works)*

The prototype shop produces at feeds and speeds which are best in a low-volume operation with a minimum tool cost as the desired result. A higher-volume production shop operates at feeds and speeds engineered for maximum productivity. Each operation requires a different chip-cutting philosophy, yet each may be cutting similar parts made of the same material and manufactured on the same kind of N/C machine. Such

Fig. 1–5.—Three-axis, three-spindle "bridge-type" N/C contouring machine. *(Courtesy, Cincinnati Milacron, Inc.)*

Fig. 1-6.—Adjustable cutting-tool holders. Differences in flexibility of such tool holders and cutting tools may require varied machinability data. *(Courtesy, Microbore Division, DeVlieg Machine Company)*

conditions may even exist within the same shop because of the inherent flexibility of N/C operations. The need for specialized data for N/C must be recognized in order to solve many of the unique tool-life problems of N/C manufacturing.

Producibility. Producibility is the relative ease or difficulty with which a part can be produced. Producibility of a given part is often improved by the use of N/C machine tools, not only because the machines can be operated at higher speeds and feeds than conventional tools, but because multiple-spindle N/C machines can produce more than one part at the same time (6). Specialized machinability data is a necessity to maintain the highest possible rate of production under N/C producibility conditions.

Changes in Responsibility

The final reason that specialized data is needed for N/C operations is that N/C has caused a major shift in the function of the machine-tool operator. Whereas the operator of a conventional tool must be skilled in the techniques of material removal, with some knowledge of machinability parameters, today's N/C operator is usually responsible only for loading and unloading the machine and starting, stopping, and monitoring the progress of the program. Vital as these functions are, a knowledge of machining data is not a requirement of an N/C machine-tool operator. The functions of determining feeds, speeds, tool life, and other machining parameters are now the responsibility of other men—the manufacturing engineer, the process engineer, the tool analyst, and the N/C parts programmer. Each of these men must be supplied with the specialized data he needs to perform his part of the total preparation for N/C.

Manufacturing Engineer. The job of the manufacturing engineer is one of the most important to efficient N/C production. His responsibilities are overall development of the process sequence of operations, establishment of tooling parameters, and preparation of the parts program. He must consider part characteristics, work content, machine capability, cycle time, lead time, order quantity, availability of tooling, and the economics of N/C machines.

Process Engineer. The process engineer is specifically responsible for the N/C machining process, the sequence of machining operations that will be performed on the part. The process he develops will also establish the specific type of equipment to be used and will usually dictate tooling requirements.

Tool Analyst. The tool analyst coordinates his work between the process engineer and the parts programmer. He establishes cutting-tool parameters, and their associated machinability data, based on the requirements of the process. He must select cutting tools with the proper geometry, methods of holding the cutting tools, and cutting-tool preset dimensions, if required.

Parts Programmer. The parts programmer's job is to develop the process which converts the product design into a set of instructions which can be recognized, interpreted, and carried out by the electronic control system of an N/C machine tool. In some facilities, part programming is carried out with the assistance of a computer. In other locations the task is accomplished manually. In either case the programmer initiates and controls the activity.

In some N/C operations, programming is a team effort; in other facilities one individual accomplishes all the work. Regardless of the mode of programming, a finished part from an N/C machine tool reflects the coordination and concern for detail that must be a part of the programming process. The automatic operation of N/C equipment demands discipline which is not often developed in conventional operations. An N/C job cannot come to a successful conclusion unless all details are accomplished accurately and are presented in time for use. Specialized machinability data is needed for successful programming.

MACHINABILITY DATA INTERRELATIONSHIPS

The use of machinability data in the total N/C concept requires coordination from product design to machine-tool operation. Each characteristic of the part—material, size, shape, strength, tolerance, etc.—and each machining operation used has an impact on the use of machinability data. Each of these characteristics should be considered as the tooling and the program for the process method are developed. In addition, the manufacturing engineer and the parts programmer must relate the characteristics of the part to other aspects of production, such as the total number of parts to be made, lot sizes, lead time, and available machine time, in order to find optimum machining efficiency—that level of operation which results in minimum cost or maximum production.

Part Process

The conditions established by the part process are interrelated with all the factors mentioned above because the part process contains the sequence of machining operations. This sequence is established to permit manufacturing under the optimum conditions allowed by the manufacturing philosophy for the product.

The machining sequence is primarily a reflection of part design. Many parts selected for N/C machining require a combination of different types of machining operations. The sequence of operations to be used determines part accuracy by removing metal in planned stages under stringently controlled conditions. For example, roughing, semifinishing, and finishing operations are designed to control the size and finish of a part.

The same operations may also be intended to reduce the effects of heat distortion in a part. On conventional machines, each of these operations is accomplished in a different setup. As one operation is completed, the part is set aside to await the next step of the process, and a "cooling-off" period is consequently provided between setups. This advantage is not provided in a lengthy N/C operation on a complex part, and the sequence of machining operations and the associated machining data used require a new process of thought. To produce a quality product, N/C machining data and a logical sequence of machining operations must be devised which will avoid overheating the part.

Fig. 1–7.—Horizontal boring, drilling, and milling machine. *(Courtesy, The Cincinnati Gilbert Machine Tool Company)*

Fig. 1–8.—Precision positioning and contouring machining center. *(Courtesy, Kearney and Trecker Corporation)*

Fixturing

Another important relationship of machinability data is in the area of fixture design, which, for N/C applications, encompasses quite a broad range. Depending on the number of parts to be produced, N/C fixtures may be developed from universal modular components (so-called free-setup types) or from relatively simple holding fixtures. Often highly specialized fixtures must be used. Each method has its own pertinent set of parameters which will affect the use of machining data. Sometimes the combined effects of a fixture structure and a highly flexible N/C machine permit higher speeds and heavier cuts than would normally be used. At other times the nature of the machine and the operations to be performed will demand a compromise in fixture design. Fixture strength and rigidity may have to be sacrificed in order to accomplish more operations within one setup. Also, in some cases, the universal modular type of fixture lacks rigidity or necessitates longer cutting tools to avoid interference or collisions. Machining data should be used as a guide to the most effective fixture design possible under the circumstances of production.

Cutting-Tool Control

The control of cutting tools must also be related to machinability data for N/C machining. Cutting tools for N/C are required to perform specific tasks under given conditions, and the right types of cutting tools, made of the right types of material for the application and ground to the particular geometry to complete the task, must be supplied. Many types of standard tools are marketed in several standard geometries for a given job, and often the only sure way of selecting the proper geometry is on the basis of test results. The effect of tool geometry on tool life, surface finish, and surface integrity is highly complex, and specific data are needed before the proper selection can be made. A tooling system for N/C should be set up to provide the specified tool, when it is called for, with the necessary parameters about which the part program is designed.

For example, Fig. 1–9 shows tap life vs. tap design for four different types of taps as used on a particular austenitic stainless steel. All the taps are standard off-the-shelf items. If the three-flute tap were to be tried first, the material might be declared untappable. Machining time could be saved by testing all the taps, but if a chart like the one shown or other comparative data on the four taps were available, the programmer or tool analyst could specify the two-flute, plug, chip-driver tap and avoid a great deal of experimentation and lost time (5).

Similarly, the effect of side cutting-edge angle on tool life is shown in Fig. 1–10. This geometry may or may not fit a particular machining situation, but if the programmer or tool analyst does not have this information available, and does not know that such a relationship exists, then he cannot be expected to make a choice of cutting tools based on the facts (5).

Machine Tool Operator

The last person involved in the long chain of N/C control, but certainly not the least important, is the N/C machine-tool operator. This individual is too often over-

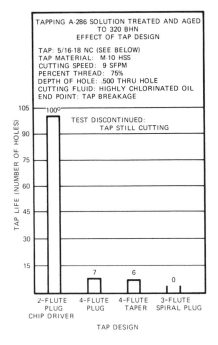

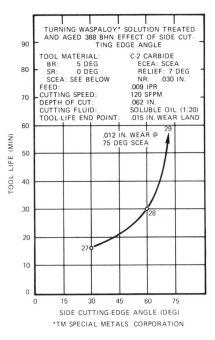

Fig. 1–9.—Tap life vs. tap design for A-286 stainless steel (7).

Fig. 1–10.—Effect of side cutting-edge angle on tool life for turning (8).

looked during the planning of N/C operations. Because of the magnitude of intelligence contained in the coded functions of the control tape, the operator may feel that he is not responsible for the functions of the machine. He should be made aware of his responsibilities for each part through a set of operating instructions which spell out clearly and in detail what he is expected to do. The doors of communication should always be left open to him as well. Because of his constant attention at the machine tool, the operator can complete the circuit of a closed loop by feeding back information which may assist in developing specialized data systems and in optimizing the operation of N/C machine tools.

REFERENCES

1. A. A. Ameday, "Management Systems for Machine Tool Utilization,"*ASTME Paper No. MM68-702* (1968).
2. P. J. Anderson and P. F. Boyer, "Effective Utilization of Numerically Controlled Turning Machines at the Oak Ridge Y-12 Plant," *SME Paper No. MR70-199* (1970).
3. Terry J. Fischel, "N/C Machine Utilization Monitoring System," *SME Paper No. MR70-200* (1970).
4. Frank E. Smith, "Automatic Clamping as It Affects the Utilization of N/C Machines," *SME Paper No. MR70-201* (1970).
5. Roy L. Williams and William B. Johnson, "Applicability of Machining Data to Numerical Control," *ASTME Paper No. MR68-711* (1968).
6. Raymond E. Howe, ed., *Introduction to Numerical Control in Manufacturing,* Dearborn, Mich., American Society of Tool and Manufacturing Engineers, 1969.
7. *U.S. Air Force Machinability Report,* Vol. 4, "Machining Characteristics of High Strength Thermal Materials (AMC-TR-60-7-532)," Cincinnati, Air Force Machinability Data Center, 1967.
8. *Final Report on Machinability of Materials (AFML-TR-65-444),* Vol. 6, Cincinnati, Metcut Research Associates, Inc., 1966.

CHAPTER 2

TYPICAL DATA SYSTEMS FOR N/C PROGRAMMING

Theodore Judson, General Motors Institute
The first chapter has given general background in the development of machinability data for N/C machining, some of the reasons such data is needed, and some of its relationships with the machining processes. This chapter details typical machinability data systems that are in use today, their preparation, and their use in program development.

DATA AND DATA FORMATS

In any N/C system, machinability data, and other related information about tools or materials, must be converted into a form that can be read and operated on by an N/C machine-tool control system. Even with the most simple, manual methods of control preparation, some type of data-handling machine must be used to make that conversion. The data-handling machine may be as simple as a modified commercial electric typewriter or as sophisticated as a modern computer. In either case, *data* may be defined as that information that is entered into, manipulated by, and recovered from the data-handling machine in order to prepare the facts needed for N/C control.

A *programming system*, or *program*, is a set of instructions that defines the conditions for a desired N/C process and the operations needed to meet those conditions. When a programming system is to be manipulated by computer, it is referred to as *software*. The sources of N/C control data stem directly from programming system requirements as well as workpiece requirements. An analysis of workpiece specifications provides data on size, shape, material, and the work content involved in processing the part. At the same time, programming systems have their own unique language and information requirements that must be added to the control data. For example, tool descriptions and machining parameters must be included in the programming system in order to resolve any processing problem. No matter how sophisticated a programming system might be, data concerning machining conditions, the N/C machine tool, the machine control system, the communicating language, and the workpiece must be combined in the data-handling machine to produce an N/C machine control tape.

There are both special-purpose and general (or universal) programming systems. One special-purpose program is called SPLIT, an acronym for *S*undstrand *P*rocessing

15

*L*anguage *I*nternally *T*ranslated. This program is designed for two-axis positioning with some contouring capability on a particular series of machine tools. One of the several universal programs is APT—*A*utomatically *P*rogrammed *T*ools. APT is highly complex and sophisticated, and it provides multiaxis machining capability to many types of machines. Another typical software system for computer-aided data handling is shown in Fig. 2–1. The software in this system, called ADAPT/AUTOSPOT (*AD*-aptation of *A*utomatically *P*rogrammed *T*ools/*AUTO*matic *S*ystem for *PO*sitioning

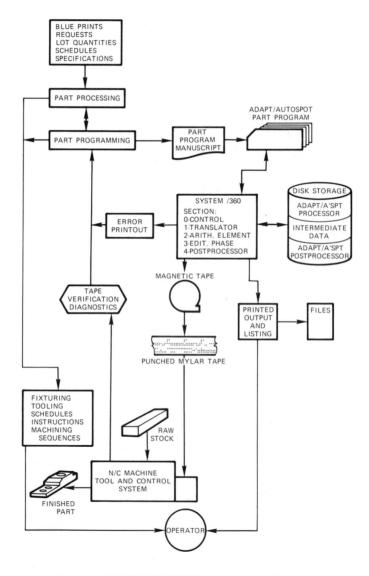

Fig. 2–1.—ADAPT/AUTOSPOT software system flow chart.

Tools), is designed to allow combined or separate position and contour programming through small-scale computers.

Data Format

A *data format* may be defined as the general order or arrangement in which information is manipulated by a data-handling machine to produce an N/C machine control tape. Information such as machining conditions, coordinate axis locations, preparatory and miscellaneous functions, and geometric workpiece configurations are specified by the parts programmer and are entered in a designated format on a work sheet commonly called a coding form or programming manuscript. The information, in the proper format, is then entered into the data-handling machine, which converts the data into the coding on an Electronic Industries Association (EIA) standard eight-channel, one-inch-wide, punched machine control tape.

Format, as related to the machine-tool input medium, is also the general physical arrangement in which data groups appear on the control tape or on control cards. These data groups are made up of blocks, which are in turn composed of words, A *block* is the data which will provide complete information for one unit of machine-tool operation—movement of the tool from one contour point to another, for example. A *word* is the smallest unit in which meaningful data may be written. Data describing a

CINCINNATI MILACRON — Cintimatic Division
Program Sheets
Cintimatic Vertical and Horizontal Machining Centers

PART NAME: NO. 3 CINTIMATIC TEST BLOCK
SET UP & TOOL INFO.: POSITION NO. 30: X–00.625, Y–05.500; POSITION NO. 31: X–02.500, Y–06.625; POSITION NO. 32: X–09.500, Y–06.625. USE ¾" PIN AS SHOWN ON COORD. DWG. 706620.
DRAWING No.
FIXTURE No.
PART No. 706620
PAGE 1 OF 3 PAGES
PROGRAMED BY: A. ABLE
DATE: 12/20/70

POS No.	H/O or N SEQ #	G PREP FUNCT	X POSITION	Y POSITION	I POSITION	J POSITION	F FEED RATE	W CAM	MISC FUNCT	TUR STOP	SPDL FD	RPM	DEPTH OF CUT	REMARKS
21	N001	G80	X00,687	Y07,625					M51			750	375	5/8 DIA. E.M. (SET Z DEPTH FROM SURFACE A) USE TURRET STOP NO. 1 FOR CAM M51-M55
21	N002	G79												
22	N003			Y00,687										
23	N004		X12,125											
23	N005	G81												
25	N006	G78	X07,812	Y002,250					M52					SET Z DEPTH FROM SURFACE C. (CLAMP QUILL) UNCLAMP QUILL--RETRACT
27	N007	G79		Y00,625										
27	N008	G78												
26	N009	G80	X03,937	Y07,625					M53					SET Z DEPTH FROM SURFACE A
26	N010	G79												
24	N011			Y00,250										
24	N012	G81												
18	N013	G80	X10,250	Y01,437					M52					
18	N014	G79												
17	N015			Y02,875										
35	N016		X07,937											
35	N017	G81												
35	N018	G79	X07,937	Y02,875					M54					SET Z DEPTH FROM SURFACE B
34	N019		X03,812											
34	N020	G81												
34	N021	G79	X03,812	Y02,875					M53					SURFACE A
29	N022		X03,250											
29	N023	G81												
19	N024	G80	X01,687	Y04,875					M53					
19	N025	G79												

Fig. 2–2.—Typical format and data for a two-axis N/C machine. Data for speed, feed, tool description, and tool change are not included as part of the machine control tape. *(Courtesy, Cincinnati Milacron, Inc., 5)*

single movement of the tool along a single coordinate axis is an example of a data word. Three formats in current use for machine-tool input media are the following:

1) Word-Address Format—In a word-address format, each data word in a block is identified by a specific character, usually alphabetic, which establishes the word's meaning.

2) Tab Sequential Format—In a tab sequential format, words are separated by tab code characters. The location of the word in the block establishes the word's meaning. For example, X-coordinate words always precede Y-coordinate words.

3) Variable-Block Format—A variable-block format allows the number of words in each block to vary from block to block. The blocks may be word addressed but of variable lengths.

Other formats are used, and they too must be understood by the parts programmer before he can communicate information to a data-handling machine and produce a suitable control tape for a specific N/C machine tool. These variations in format are real and important, but are not difficult to work with. N/C machine-tool builders select the formats best suited to fit their machines' control systems and capabilities. Each format selected remains uniform and compatible to the entire programming system whether that system is manual or computer-aided. The format shown in Fig. 2–2 is the word-address type, and the block of data given is typical for the operation of a single-spindle, two-axis N/C machine (5).

Formats used by N/C parts programmers serve as organized communications devices. Although formats may vary in form and content, they provide N/C users with a common ground for understanding the language and data contained in any input medium. The parts programmer is not required to understand how or why software operations or data manipulations work, nor is he required to understand the internal workings of the computer. He must merely know how to communicate with the particular type of N/C equipment in his own manufacturing installation.

Formats and the Process Sequence

The sequence of events necessary to produce an N/C machine control tape is not unlike the steps required to plan the manufacture of a part by conventional techniques. As shown in Fig. 2–1, product drawings are analyzed to select the part process. The process selected may be based upon such information as production quantities required, workpiece material, production schedules, and workpiece specifications. At this stage in planning, a machining sequence is established, and it is followed by the selection of tools, the design and fabrication of jigs, fixtures, and gages, and the procurement of suitable production machines.

Following part processing, N/C part programming analysis is begun. When this step is completed, a part program manuscript containing workpiece and process requirements in tape or card form is prepared and combined with the part programming system to produce an N/C machine control tape. Except for the programming function and format, much of the paperwork used up to this point are very similar for N/C and conventional machining methods. Figs. 2–3 and 2–4 illustrate a routing sheet and a part drawing used by one leading N/C builder to bring a workpiece to final part specifications (6).

The data format provides an important record of pertinent manufacturing details. If it is revised when engineering changes are made, this record can establish correct setup points, speeds, feeds, and tool types and configurations, and it can represent a success-

PROCESS MASTER FOR NAS 913, REV. 2, AUG. 15, 1965 TEST 4.3.3.8.5		3-AXIS PROFILER
BENDIX INDUSTRIAL CONTROLS DIVISION 12843 GREENFIELD ROAD DETROIT, MICHIGAN 48227	BICD J.O. CUST. P.O.	MACHINE TOOL BUILDER END USER
PROCESSED BY	APPROVED BY	DATE

OP.	DESCRIPTION	
10	SECURE AND PREPARE STOCK AS PER SKETCH #10.	MINIMUM PREPARATION TO CONFORM TO SPECIFICATION
	IF SUGGESTED FIXTURE PLATE, SKETCH #13, IS TO BE USED ADDITIONAL PREPARATION AS PER SKETCH #12 IS REQUIRED.	
15	SET UP STOCK ON MACHINE TABLE AS PER SKETCH #15.	REQUIRED BY SPECIFICATION
20	CUTTER WILL RAPID FROM SET UP POINT PAST EDGE OF STOCK, PLUNGE, ROUGH CUT DIAMOND, AND RETURN TO SET UP. SEE SKETCH #20.	
30	CUTTER WILL RAPID FROM SET UP PAST WORK PIECE, PLUNGE, ROUGH CUT CIRCLE AND RETURN TO SET UP. SKETCH #30.	OPSTOP'S ARE PROGRAMMED
40	RAPID FROM SET UP, PLUNGE, ROUGH CUT SQUARE AND RAMPS, AND RETURN. SKETCH #40.	.030 STOCK IS LEFT BY ALL ROUGH CUTS.
50	RAPID FROM SET UP, PLUNGE, FINISH CUT SQUARE AND RAMPS, AND RETURN TO SET UP.	
60	CUTTER WILL RAPID FROM SET UP, PLUNGE, FINISH CUT THE 4 OUTER EDGES OF THE CIRCLE BASE PLANE, AND RETURN	
70	RAPID FROM SET UP PLUNGE, FINISH CUT THE CIRCLE AND RETURN TO SET UP.	

Fig. 2-3.—Portion of a descriptive operation and routing sheet for a three-axis N/C contouring profiler. (*Courtesy, Bendix Industrial Controls Division, 6*)

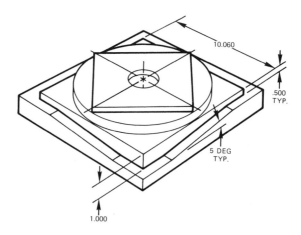

Fig. 2-4.—Drawing of the part corresponding to Operation 40 in the descriptive routing of Fig. 2-3 (6).

ful approach and definite solution to a particular machining problem. This information can and should be used as a means of developing manufacturing standards and procedures for similar machining problems. It is a source of proven machinability data.

Format Sophistication

Format sophistication is directly related to the complexity of the N/C machine tool. As the level of sophistication of software and N/C machine tools increases, the complexity of input data also increases. The amount of data entered in each block of tape increases as N/C machines are developed which allow more control functions, and it may become necessary for the parts programmer to enter data in the format to control table indexing, tool changes, and speeds and feeds in addition to coordinate axis positions.

It is true that a three-axis N/C contouring machine; characterized by a highly complex control system, can perform operations usually assigned to a less complex N/C machine such as a two-axis drill. This situation, however, is inefficient from a cost standpoint because an expensive machine, designed for complex work, is working at less than its maximum capability. As N/C machines and control systems increase in sophistication, the analysis of workpiece specifications is more involved and the software and data required become more complex.

It also becomes increasingly important that the parts programmer specify the best and most accurate machinability data available for entry into the system. Only by using reliable machinability information can tool life and surface finish be controlled and production time be maximized. The total system—N/C machine, software, and computing system—is expensive to operate and maintain. Nonproductive time is costly. The following examples illustrate the relationship of machining data to machine complexity by considering the control capabilities and the programming system requirements of several N/C machine tools.

Two-Axis Positioning. A two-axis positioning machine designed for drilling, tapping, boring, and light milling is a relatively simple machine. Its control system format may be of the word-address type. Essentially, the machine is capable of N/C control in the X and Y coordinate axes but requires manual speed, feed, and depth settings. A simple programming system may be used, and tape preparation may be done manually. If workpiece characteristics are complex, computer-aided programming may be more efficient. Such two-axis positioning machine control systems respond to the following input data:

1) Process sequence number coding
2) Preparatory functions (predetermined machining cycles)
3) X and Y coordinate position information
4) Miscellaneous function codes (predetermined functions other than machining cycles).

In addition to supplying this controlling information, the part program should also specify setup positions, establish the speed, feed, and depth of cut for each operation, and provide a tool description; but none of this information is necessary in, or appears in, a block of words on the machine control tape. The format and data for a simple two-axis N/C positioning machine are shown in Fig. 2-2 (5).

Three-Axis Positioning. By increasing the complexity of the machine and the control system, a machine can be designed which will drill, bore, tap, and mill with positioning control in the X, Y, and Z axes, and which in addition may provide tool changes, pallet shuttling, table indexing, and complete programming of speeds, feeds, and depths of cut. Some machines of this type have control systems which enable them to perform approximate angular and circular motions. This feature is useful, but it is not suitable for accurate profile and contour machining. System programming may be accomplished manually, but often computer-aided methods are found to be efficient in creating the control tape. The input data format may be of the tab sequential type. A three-axis positioning machine control system responds to the following data:

1) X, Y, and Z coordinate axis position information
2) Speed, feed, and depth of cut information
3) Tool selection instructions
4) Auxiliary or miscellaneous function codes
5) Table index position descriptions
6) Automatic pallet shuttle directions
7) Slope and arc cutting instructions.

The parts programmer will also provide setup position information, operation sequence numbers and descriptions, and important tool description information such as type, overall length, effective length, and point angle. All this information may be used to create a machine control tape by manual methods. If computer methods are employed, however, the parts programmer may select the ADAPT/AUTOSPOT software illustrated in Fig. 2-1. The machine described in the preceding paragraphs is a positioning machine and makes use of only the positioning (AUTOSPOT) capabilities of this program. The need for computer aid may be based upon part specifications in which patterns, translations, inversions, and other complexities are encountered.

The use of the AUTOSPOT portion of the programming system is based upon a symbolic program system (SPS), and the programmer enters statements in a symbolically addressed, variable-length sentence form in three general sections: (1) a definition section which is concerned with the machine tool and part, (2) a tool information section which describes the required tools, and (3) a machining section which describes pertinent information about a specific machining requirement. All data is written, as

Part Name: SAMPLE ARC & SLOPE PROGRAM	Drawing Date	Material	Page 1 of 2	Operation No.	Part No.

Seq. No.	Operation and Tool	Longitudinal (X)	Vertical (Y)	Depth (Z)	Feed Rate	Spindle Speed	Tool Code Number	Aux. Function	Table Index	Start	End	Radius	Dis.	Time
1	POSITION X, Y, & Z TO POINT 1	074 603	050000	08750		0400								
2	FEED X TO POINT 2 SET START FOR ARC 2-3	130000			12	0400				000				
3	CUT ARC 2-3	144140	055860		12	0400					125	2.0	1	
4	SET START FOR SLOPE 3-4					0400								
5	CUT SLOPE 3-4	174140	085860		12	0400		11						
6	SET START FOR ARC 4-5					0400				125				
7	CUT ARC 4-5	180000	100000		12	0400					250	2.0	1	
8	FEED Y TO POINT 6 SET START FOR ARC 6-7		130000		12	0400				250				
9	CUT ARC 6-7	160000	150000		12	0400					500	2.0	1	
10	SET START FOR ARC 7-8					0400				500				
11	CUT ARC 7-8	145 860	144140		12	0400					625	2.0	1	
12	SET START FOR SLOPE 8-9					0400				625				
13	CUT SLOPE 8-9	105860	104140		12	0400		11						
14	SET START FOR SLOPE 9-1					0400				666				
15	CUT SLOPE 9-1	074603	050000		12	0400		11						

Fig. 2–5.—Format and data of a slope and arc program for a three-axis-positioning N/C machining center with auxiliary equipment (7).

in the manual method, on a format sheet and is then punched on computer cards. When combined with the ADAPT/AUTOSPOT general program processor and a suitable post-processor for the machine tool, an N/C machine-tool control tape, to EIA standards, is created. The format used, and the data entered into it by the parts programmer, is illustrated in Fig. 2–5 (7).

Contouring. Contouring may be performed with single or multiple spindles on multiple-axis N/C contouring machines. A three-axis control, single-spindle machine tool is typical of this machine configuration. Accessory capabilities such as tool changing, rotary table indexing, and pallet shuttling may be part of the machine configuration. The machine control system usually is capable of simultaneous motion in all axes to provide for continuous motion control of the cutter path. Analog methods may be employed to create and monitor a two- or three-dimensional cutter path by linear, circular, or parabolic interpolation.

The data input and format for contouring must adhere to one of several programming systems and languages presently available. One such program is the IBM System/360 (S/360) version of ADAPT and is a portion of the total programming system shown in Fig. 2–1. ADAPT is limited to sophisticated, two-dimensional contouring in the X-Y plane or an inclined plane. The language, syntax, and format of ADAPT are almost identical to those of the APT system. The intricacies of this format and its data are shown in Figs. 2–11 through 2–13. The parts programmer may specify the following data in a language similar to FORTRAN:

Geometric Statements. Geometric statements describe the following geometric figures:

1) Point (eight methods of description)
2) Line (twelve methods)
3) Circle (eight methods)
4) Tabulated cylinder (two methods)
5) Plane (five methods)
6) Vector (nine methods)
7) Matrix (five methods).

Motion Statements. Motion statements describe the following:

1) Specifications of drive, part, and check surfaces
2) Tool movement specifications in relation to the part (tool left, tool right)
3) General motion instructions (go forward, go back, go left)
4) Start-up and termination commands.

Computing Statements. Computing statements consist of the following:

1) FORTRAN-like statements for computations similar to those of FORTRAN
2) Macros for repetitive programming.

Special Statements. Other statements for special purposes include the following:

1) Machine-tool start-up and termination commands
2) Cutter definitions
3) Tolerance, feed rate, and other machining parameter specifications
4) Input-output, control, and debugging statements.

MACHINABILITY DATA ACQUISITION

Sources of accurate and applicable machinability data are necessary for success in N/C machining. This does not imply that the data used for N/C processing is any more accurate or in any way better than the information needed and used for other production machining methods. This does mean, however, that in its objectives and philosophy N/C manufacturing represents a high level of automatic control and therefore requires accurate data.

The man/machine relationship of manufacturing has evolved from one of interference with, decision making about, and adjustment of the manufacturing process to manufacturing in which predictions about the success of the operation are used in a program to control that operation without man as an overseer. Realistically, this N/C philosophy has not yet been attained simply because both N/C and conventional process techniques are subjected to the same operational disturbances, some of which are (1) tool wear, (2) deflection, (3) workpiece variation, and (4) thermal variation. Until accurate, reliable, and predictable control can be established over such variables, machinability data will, by necessity, remain largely general in nature. On a workpiece that is unique in material and configuration, both N/C and conventional processes must use sources of machinability data that are essentially of the same reliability.

Machinability data is basically that information related to a prediction about the life of a cutting tool as compared to some known reference point. Such data must include all the variables which interact and affect machining performance. In broad groupings, these variables may be listed as follows:

1) Cutting-tool designs and materials
2) Tooling characteristics

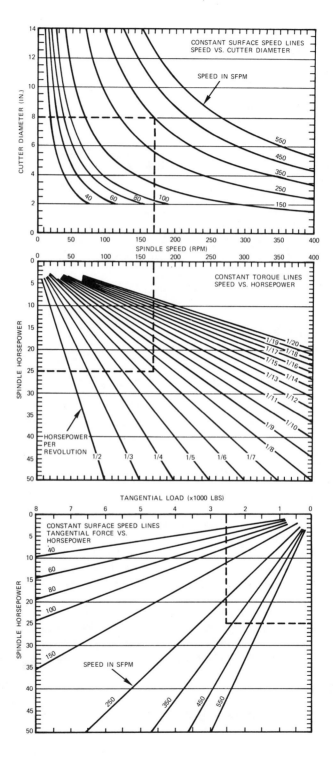

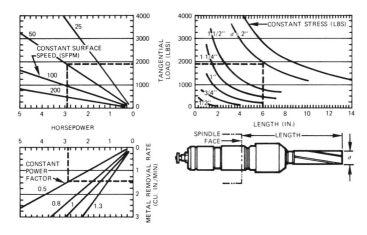

Fig. 2–6.—*a*, opposite page, nomographs for surface speed, torque, and tangential load for milling; *b*, above, nomographs for tangential load, horsepower, and metal-removal rate for milling. *(Courtesy, Sundstrand Corporation and* Machinery, 8*)*

   3) Workpiece properties
   4) Machine-tool characteristics
   5) Dimensions of the cut
   6) Cutting fluids.

It is within these groupings that the operational disturbances listed above affect the data used to achieve predictable surface finish or tool life.

Data Categories

The types of data used by the parts programmer are classified in three categories. It is the programmer's responsibility to combine and effectively use the information available in each category.

Workpiece Material Data. The first category of data used by the parts programmer concerns workpiece materials. These data are material-to-material comparisons usually based upon a relationship of tool life to machining speed or upon the achievement of a required surface finish. The kind of data in this category, and its sources, will be discussed later in this chapter.

Cutting Data. The second data group may be designated by the cutting tool or the type of operation used. The data may be classified for turning, drilling, milling, or any other general operation and may be designed to show how variations in tool geometry and machine settings affect cutting loads, cutter deflection, and metal-removal rates as important factors in good programming. The programmer may find this type of machinability data valuable in estimating minimum machine horsepower requirements and maximum practical cutting-tool load levels. The use of nomographs such as those shown in Fig. 2–6 may aid the programmer in increasing metal-cutting productivity and may strongly influence the success or failure of N/C operations.

The nomographs in Fig. 2–6*a* may be used to determine surface speed, torque, and tangential load for milling operations. The example shown by the heavy dashed lines is for an 8-in.-diameter cutter run at 170 rpm in a 25-hp machine. In the first nomograph, the cutter-diameter and spindle-speed lines meet at the 350 sfpm curve. Plotting spindle speed and horsepower on the next nomograph shows that torque is 1/7 horse-power per revolution. Finally, plotting spindle horsepower against surface speed (sfpm) in the last nomograph indicates that the tangential load is approximately 2,400 lbs. If torque and tangential loads exceed the design limits of the machine and the cutting tool, breakage may occur.

In Fig. 2–6b the nomographs show tangential load, horsepower, and metal-removal rate for milling. The tool drawing indicates the combined length of cutter and holder. (If flange-type holders are used, only the length of the cutter applies.) The example shown by the heavy dashed lines is for a 2-in.-diameter cutter that extends 6 in. from a flange-type holder. The length line is plotted in the first nomograph so that it intersects the correct constant stress (cutter-diameter) line. (If the cutter's shank diameter is smaller than its largest diameter, the shank diameter should be used.) Tangential load at the cutting end of this example tool is approximately 1,900 lbs. The second nomo-graph is used to find horsepower, and the third is used to determine material-removal rate. The power factors for various materials may be found in Table II–1. Here, these factors are applied only to face mills and must be multiplied by .55 for use with end mills (8).

Machine-Tool Cutting Data. The third category of machinability data is cutting data based on a particular machine tool's capabilities. This type of data, illustrated in Tables II–2 through II–9, is commonly generated by the machine-tool builder, usually in response to a specific materials cutting problem. This data will reflect machine and cutting-tool operating parameters which have been found successful in machining a given group of materials under controlled conditions. The objectives may be machine oriented, but the results remain similar to those of other categories of machinability—achievement of an equitable tool life, a controlled surface finish, and the optimum use of horsepower.

Data Sources

Machinability data is generated from several sources. In its published form, the large body of such information is the direct result of programs specifically designed to create machinability data. These programs may be conducted in the laboratories of colleges and universities or in the research and development departments of industry. The small plant finds it more convenient and necessary to use data collected from exter-nal sources, for it usually has neither the personnel, the time, nor the finances required to generate elaborate internal data-collection systems. Some published standards exist and may be used by small organizations to their advantage, but the great bulk of their process knowledge will be formed through individual experience supplemented by data published by others.

Most large manufacturing enterprises develop programs designed to create and expand machinability data. Many develop operating standards from intensive machina-bility data studies carried out "in house" or under financial grants to colleges, universi-ties, or research firms. Their objective remains the same, however—to obtain reliable machinability data for use in the manufacturing process.

There are three major sources of machinability data—personal experience, tool builders' specifications, and handbooks.

Personal Experience. Much machinability data is gained from experience in production. It may not be published, but may rather be simply carried in the memories of key personnel and applied as needed. This is not an ideal situation, because the loss

Table II-1. Average Unit Power Requirements (9).

Material	Hardness (BHN)	Unit Power* (hp/cu. in./min)					
		Turning HSS & Carbide Tools Feed 0.55–.015 ipr		Drilling HSS Drills Feed .002–.008 ipr		Face Milling HSS & Carbide Tools Feed .005–.012 ipr	
		Sharp Tool	Dull Tool	Sharp Tool	Dull Tool	Sharp Tool	Dull Tool
Steels—Wrought	85–200	1.1	1.4	1.0	1.3	1.1	1.4
and Cast	35–40 R_c	1.4	1.7	1.4	1.7	1.5	1.9
Plain carbon	40–60 R_c	1.5	1.9	1.7	2.1	1.8	2.2
Alloy steels	50–55 R_c	2.0	2.5	2.1	2.6	2.1	2.6
Tool steels	55–58 R_c	3.4	4.2	2.6	3.2†	2.6	3.2
Cast Irons							
Gray, ductile, and	110–190	0.7	0.9	1.0	1.2	0.6	0.8
malleable	190–320	1.4	1.7	1.6	2.0	1.1	1.4
Stainless Steels							
Wrought and Cast	135–275	1.3	1.6	1.1	1.4	1.4	1.7
Ferritic, austenitic							
and martensitic	30–45 R_c	1.4	1.7	1.2	1.5	1.5	1.9
Precipitation Harden-							
ing Stainless							
Steels	170–450	1.4	1.7	1.2	1.5	1.5	1.9
Titanium	250–375	1.2	1.5	1.1	1.4	1.1	1.4
High Temperature							
Alloys, Nickel and							
Cobalt Based	200–360	2.5	3.1	2.0	2.5	2.0	2.5
Iron Based	190–320	1.6	2.0	1.2	1.5	1.6	2.0
Refractory Alloys							
Tungsten	321	2.8	3.5	2.6	3.3†	2.9	3.6
Molybdenum	229	2.0	2.5	1.6	2.0	1.6	2.0
Columbium	217	1.7	2.1	1.4	1.7	1.5	1.9
Tantalum	210	2.8	3.5	2.1	2.6	2.0	2.5
Nickel Alloys	80–360	2.0	2.5	1.8	2.2	1.9	2.4
Aluminum Alloy_	30–150 500 kg	0.25	0.3	0.16	0.2	0.32	0.4
Magnesium Alloys	40–90 500 kg	0.16	0.2	0.16	0.2	0.16	0.2
Copper	80 R_b	1.0	1.2	0.9	1.1	1.0	1.2
Copper Alloys	20–80 R_b	0.64	0.8	0.48	0.6	0.64	0.8
	80–100 R_b	1.0	1.2	0.8	1.0	1.0	1.2

*Power requirements at spindle drive motor, corrected for 80 percent spindle drive efficiency.
†Carbide

Table II-2a. Classification of Materials (Steels).

S.A.E. Number	Process	Heat Treatment	BHN	Machinability Classification
Carbon Steels				
1010	CD	——	141	5A
1010	HR	——	110	3A
1015	CD	——	150	5A
1015	HR	——	143	3A
1018	CD	——	160–180	5A
1018	HR	——	120–140	4A
1020	CD	——	149–170	5A
1020	HR	——	110–140	4A
1022	CD	——	156	5A
1022	HR	——	126	4A
1025	CD	——	168	4A
1025	HR	——	160	4A
1030	CD	——	180	4A
1030	HR	——	149	4A
1030	HT	——	192	4A
1035	CD	——	170–202	4A
1035	HR	——	143–182	4A
1040	CD	——	195	4A
1040	HR	——	179	4A
1040	HT	Water Quenched	235	3A
1045	CD	——	183–228	3A
1045	CD	Annealed	——	4A
1045	HR	——	156–202	3A
1045	HT	Water Quenched	197–248	3A
1055	CD	Annealed	——	3A
1065	CD	Annealed	——	2A
1080	CD	Annealed	——	2A
1095	CD	Annealed	——	2A
1095	HT	——	300	1A
1042	CD	——	——	4A
Free Cutting Steels				
1112	CD	——	——	8A
1112	HR	——	126–150	6A
1113	CD	——	170–202	9A
1113	HR	——	125–150	5A
1114	CD	——	——	6A
1115	CD	——	140–170	6A
Nickel-Chromium Steels				
3141	CD	Annealed	228	4A
3141	HR	Annealed	210	3A
3141	HT	——	255	1A
3150	CD	Annealed	——	4A
3150	HR	——	230–250	3A
3150	HT	——	285–310	2A
3220	CD	——	207	4A
3220	HR	——	185	3A
3220	HT	——	225	2A
3230	CD	——	235	3A
3230	HR	——	207	2A
3230	HT	Oil Quenched	302–352	1A
3340	HR	——	288	2A
3340	HT	Oil Quenched	241–375	1A
3312		Annealed	——	3A

Table II-2a. Classification of Materials (Steels). (*Continued*)

S.A.E. Number	Process	Heat Treatment	BHN	Machinability Classification
Molybdenum Steels				
4130	CD	Annealed	187–217	4A
4130	HR	——	163–212	4A
4130	HT	——	277	3A
4140	CD	Annealed	196–228	4A
4140	HR	——	170–223	3A
4140	HT	Oil Quenched	223–363	2A
4150	CD	Annealed	——	3A
4150	HR	——	——	3A
4150 (non-sulf.)	HT and Drawn	(32–42 R_c)	——	1A
4150 sulf.		——	——	4A
4150 sulf.	HT and Drawn	(37–42 R_c)(350–400 BHN)	——	2A
4145	CD	——	——	3A
4145 sulf.	CD	——	——	4A
43B17		(Boron Steel Similar to 4320)	——	3A
4320		——	——	3A
4340	CD	Annealed	——	3A
4340	HR	Annealed	240–265	3A
4340	HT	Oil Quenched	241–415	1A
4342		Annealed	——	3A
4342	HT	——	340	1A

Table II-2b. Classification of Materials (Cast Iron).

Type	BHN	Machinability Classification
Semisteel	——	5B
Cast Steel	——	5A
Alloy Cast Steel	——	1B
CI Soft #20 Special (Z.S.C.I.)	127–150	7B
CI Medium #35 Iron and #20 Iron	161–185	5B
CI Hard #A-45 and #B-45	185–228	4B
Moly Cast Iron	——	4B
Malleable Iron	137	5B
Malleable Iron—Pearlitic	200–240 (normal K & T cstgs.)	4B
Promal	——	4B
HiNi and Cr	185	3B
HiNi and Cr Alloy	187–228	2B
No Alloy	187–228	2B
Meehanite—Gc, Ga, Gb, Gm	193–217	3B
Meehanite—Ge, Gd	174–183	4B
Hy-Test	——	4B
Hi-Alloy	165	5B
Wrought Iron	——	4B
Chilled Cast Iron	240–260	2B
Black Iron Pipe	(a soft steel approx. SAE 1020 CD)	5A

Table II-2c. Classification of Materials (Copper Alloys, Brass, Bronze).

Type	S.A.E. Number	Machinability Classification
Red Brass Casting	40	9C
Yellow Brass Casting	41	9C
Manganese Bronze Casting	43	5C
Hard Bronze Casting	62	5C
Leaded Gunmetal	63	10C
Leaded Phos. Bronze	64	9C
Phos. Bronze Gear Casting	65	5C
Cast Aluminum Bronze	68	5C
Comm. Brass Sheet	70	8C
Copper Sheet	71	3C
Tobin Bronze	73	5C
Phos. Bronze Strips	77	3C
Red Brass Sheet	79	6C
Beryllium Copper	———	5C

of an individual can result in lost time and production until the knowledge is regained by someone else. Reliance on personal experience is a rather uncommon environment in a modern manufacturing organization. Usually machinability data standards are available for reference which comprise the published work of outside investigators as well as internally generated knowledge.

Machine-Tool Builders' Specifications. The parts programmer also has the data created by the N/C machine-tool builder at his disposal. This data, usually materials oriented, is pertinent only to a particular N/C machine tool or line of tools, and it may demonstrate the machine's capability to process workpiece materials only under specific recommendations. This source of machinability information may be based upon the interests of the builder. For example, the use of N/C machine tools in the machining of new alloys is rapidly increasing. It is, therefore, to the machine builder's advantage to supply machine tools, and machinability data for their use, for the machining of these new materials.

Excellent examples of machinability data abstracted from data supplied by an N/C machine-tool builder are shown in Tables II-2 through II-9. Table II-2 illustrates general machinability classifications of materials in terms of their relative machining speeds. Various metals are first classified in major groups numbered 1 through 10. Materials in Group 1 are considered the most difficult to machine; materials in each successive group are easier to machine than the previous group. Group 8 was used as a percentage base for the first eight groups and was assigned a machinability rating of 100 percent. Groups 9 and 10 were assigned machinability percentages of 130 percent and 225 percent, respectively. The machinability ratings for the ten groups are as follows:

Material group number	Machinability rating percentage
1	30
2	40
3	50
4	60
5	70
6	80
7	90
8	100
9	130
10	225

Alphabetic designations were then assigned to various kinds of metals to show differences in general feed rates permissible in combination with machining speed ratings. The alphabetic designations are the following:

A—Steels

B—Cast irons

C—Copper and alloys, brasses, bronzes

D—Aluminum and alloys

Materials from the various groups were tested to verify their theoretical placement in the machinability range and to determine actual cutting speeds and permissible feeds at recommended speeds. The classification list, although extensive, is updated by the inclusion of new materials as they are used.

Tables II–3 through II–9 provide speed, feed, and time data for machining various material classifications. These recommendations, by the Kearney and Trecker Corporation, are for typical N/C machining operations in drilling, reaming, core-drilling, counterboring, tapping, milling, and boring. The data listed in these tables are directly related to broad materials classifications and present recommended cutting-speed and feed-rate data.

For example, assume that a manufacturer plans to drill, tap, and side mill a part from 4140 molybdenum steel that is cold drawn and annealed to a hardness of 220 BHN. Table II–2a places this material in machinability classification 4A. Information for drilling the material is shown in Table II–3 under the same machinability classification. For example, if a 1/2-in.-diameter HSS drill is to be used, the drill may be run at 535 rpm for a 70 sfpm speed and at a feed of .013 ipr. Approximate time for the drill to travel through 1 in. of the material will be .19 min.

The tapping operation for this example is described in Table II–4. Recommended speed for a 1/2-in. N.C. tap with a pitch of 13 is 229 rpm. The time to feed the tap in and out of the 1-in. through hole is .11 min. Finally, according to Table II–5, the side milling operation for a 3/8-in. depth of cut using a carbide side mill requires a speed of 237 sfpm (95 sfpm × 2.5, see note 6) and a feed of 12 ipr (8 ipr × 1.5, see note 6).

Another treatment of machinability data is used by International Business Machines Corporation for its interdivisional manufacturing. Typical data abstracted from IBM's complete standards are illustrated in Tables II–10 through II–12. These data include the following:

1) Machinability classifications and general information

2) Speeds and feeds for plain drilling

3) Speeds and feeds for milling.

Table II-3. Speeds, Feeds, and Times for Drilling,

Machinability Classification	HSS Tool Diameter	Drill Pt. and Approach	Drill 70 sfpm			Ream 40 sfpm		
			RPM	Feed	Time	RPM	Feed	Time
	1/16	.09	4280	.002	.12	2445	.010	.04
	1/8	.11	2139	.003	.16	1222	.015	.05
	3/16	.13	1426	.004	.18	815	.018	.07
	1/4	.15	1070	.006	.16	611	.027	.06
	5/16	.16	856	.008	.15	489	.033	.06
	3/8	.18	713	.010	.14	408	.036	.07
	7/16	.20	611	.010	.16	349	.039	.07
	1/2	.22	535	.013	.19	306	.039	.08
	9/16	.24	475	.013	.16	272	.041	.09
	5/8	.26	428	.013	.18	245	.041	.10
	11/16	.28	389	.013	.20	222	.041	.11
	3/4	.30	357	.016	.22	203	.041	.12
	13/16	.31	332	.016	.19	190	.041	.13
4A	7/8	.33	306	.016	.20	175	.041	.14
	15/16	.35	285	.016	.22	163	.041	.15
	1	.37	267	.018	.23	153	.041	.16
	1-1/8	.41	238	.018	.23	136	.041	.18
	1-1/4	.45	214	.018	.26	123	.041	.20
	1-3/8	.48	195	.018	.28	111	.041	.22
	1-1/2	.53	178	.018	.31	102	.041	.24
	1-5/8	.56	165	.018	.34	94	.041	.26
	1-3/4	.60	153	.018	.36	87	.041	.28
	1-7/8	.63	143	.018	.39	82	.041	.30
	2	.67	134	.018	.41	76	.041	.32
	2-1/4	.75	119	.018	.47	68	.041	.36
	2-1/2	.82	107	.018	.52	61	.041	.40
	2-3/4	.90				56	.041	.44
	3	.97				51	.041	.48
	3-1/4	1.05				47	.041	.52
	3-1/2	1.12				44	.041	.55
	3-3/4	1.20				41	.041	.59
	4	1.27				38	.041	.64
	1/16	.09	15890	.004	.02	9782	.012	.01
	1/8	.11	7945	.006	.02	4890	.018	.01
	3/16	.13	5297	.006	.03	3261	.024	.01
	1/4	.15	3972	.006	.04	2445	.027	.02
	5/16	.16	3178	.008	.04	1955	.036	.01
	3/8	.18	2648	.010	.04	1620	.039	.02
	7/16	.20	2270	.010	.04	1398	.039	.02
	1/2	.22	1986	.013	.04	1222	.039	.02
	9/16	.24	1766	.013	.04	1086	.051	.02
	5/8	.26	1589	.013	.05	979	.051	.02
	11/16	.28	1445	.013	.05	888	.051	.02
	3/4	.30	1324	.013	.06	813	.051	.02
	13/16	.31	1222	.013	.06	758	.051	.03
	7/8	.33	1135	.013	.07	701	.060	.02
10D	15/16	.35	1059	.013	.07	651	.060	.03
	1	.37	993	.013	.08	611	.060	.03

Reaming, Counterboring, and Core Drilling.

Counterbore 47 sfpm			Core Drill	
RPM	Feed	Time	Dr. Pt. and Approach	
2862	.003	.12		
1460	.005	.14		For cored holes use
962	.006	.17		75% drill speed,
713	.009	.16	.12	125% drill feed.
572	.012	.15	.13	
475	.015	.14	.15	
407	.015	.16	.16	
357	.015	.18	.17	To open drilled holes
317	.020	.16	.18	use 100% drill speed,
285	.020	.18	.20	125% drill feed.
259	.020	.19	.21	
238	.020	.21	.22	
222	.020	.23	.23	
204	.020	.25	.25	
190	.020	.26	.26	
178	.020	.28	.27	
159	.024	.26	.30	Drill point and
143	.024	.29	.32	approach includes
130	.024	.32	.35	.2 drill dia. for
119	.024	.35	.37	drill point and .07
110	.024	.38	.40	for approach and
104	.024	.40	.42	overtravel.
96	.024	.43	.45	
89	.024	.47	.47	
80	.024	.52	.52	
71	.024	.59	.57	
65	.024	.64	.62	
60	.024	.69	.67	
55	.024	.76	.72	
51	.024	.82	.77	
48	.024	.87	.82	
45	.024	.93	.87	
10180	.006	.02		
5010	.009	.02		For cored holes use
3405	.009	.03		75% drill speed,
2550	.009	.04	.12	125% drill feed.
2050	.012	.04	.13	
1700	.015	.04	.15	
1520	.015	.04	.16	
1270	.020	.04	.17	To open drilled holes
1135	.020	.04	.18	use 100% drill speed,
1020	.020	.05	.20	125% drill feed.
925	.020	.05	.21	
850	.020	.06	.22	
785	.020	.06	.23	
757	.020	.07	.25	
705	.020	.07	.26	
662	.020	.08	.27	

Table II-3. Speeds, Feeds, and Times for Drilling,

Machinability Classification	HSS Tool Diameter	Drill Pt. and Approach	Drill 70 sfpm			Ream 40 sfpm		
			RPM	Feed	Time	RPM	Feed	Time
	1-1/8	.41	883	.016	.07	544	.060	.03
	1-1/4	.45	794	.016	.08	490	.060	.03
	1-3/8	.48	722	.016	.09	445	.060	.04
	1-1/2	.53	662	.016	.09	406	.060	.04
	1-5/8	.56	611	.018	.09	374	.060	.04
	1-3/4	.60	567	.018	.10	349	.060	.05
	1-7/8	.63	530	.018	.10	326	.060	.05
	2	.67	497	.018	.11	306	.060	.05
	2-1/4	.75	441	.018	.13	272	.060	.06
	2-1/2	.82	397	.018	.14	245	.060	.07
	2-3/4	.90				222	.060	.08
	3	.97				203	.060	.08
	3-1/4	1.05				188	.060	.09
	3-1/2	1.12				174	.060	.10
	3-3/4	1.20				163	.060	.10
	4	1.27				153	.060	.11

Notes: 1. Time measured in minutes per in. of travel.
2. For drilling, core drilling, and counterboring do not increase the recommended rpm more than 5 percent to attain nearest speed on the machine. Use closest feed available on the machine.
3. Drill point and approach includes the drill point plus .07 in. for approach and overtravel.
4. Dowel-pin holes not to be bored can be drilled at basic speeds but with feeds limited to .007 in.
5. For flat-bottom drilling in holes use basic feeds and speeds.
6. For drill sizes not listed above use feed and speed for the nearest size listed. For sizes that are at the midpoint use the next larger chart size.
7. For start holes on angular surfaces or half holes use basic speed and 50 percent feed.

A thorough knowledge of how to implement machinability standards such as those illustrated in Tables II-2 through II-12 must admittedly be based on theory, practice, experimentation, and experience. The standards must be applied to a machining process by qualified personnel who can recognize improper cutting-tool design and selection and poor machining and process practices. Caution must be exercised in the direct application of data standards such as these because of differences in machining variables. The standards are intended for reference use only.

Handbooks. There are many textbooks, data handbooks, and pamphlets which provide information about the relative machinability of various materials. Usually the materials are defined in terms of their hardness, microstructure, tensile properties, and perhaps the degree of disorder given them during processing. The data give recommendations for cutting speeds, feed rates, tool materials, tool types and geometries, cutting fluids, and other general information. This kind of information is contained in such publications as the *Machining Data Handbook* (9) prepared by Metcut Research Associates, Inc., or in the booklet entitled *Speeds and Feeds for Better Turning Results* (10) by the Monarch Machine Tool Company. Examples of this type of published

Reaming, Counterboring, and Core Drilling. *(Continued)*

Counterbore 47 sfpm				Core Drill
RPM	Feed	Time	Dr. Pt. and Approach	
588	.020	.09	.30	Drill point and
529	.020	.09	.32	approach includes
482	.020	.10	.35	.2 drill dia. for
442	.020	.11	.37	drill point and .07
407	.024	.10	.40	for approach and
378	.024	.11	.42	overtravel.
353	.024	.12	.45	
331	.024	.13	.47	
294	.024	.14	.52	
265	.024	.16	.57	
231	.024	.18	.62	
214	.024	.19	.67	
192	.024	.22	.72	
183	.024	.23	.77	
170	.024	.25	.82	
159	.024	.26	.87	

8. When basic drilling feeds cannot be used because of frailty of job or fixture, use the following factors as a guide to probable performance: semirigid conditions—70 percent feed; frail conditions—50 percent feed.
9. Start drill or spot drill time = .13 min. Start drill time to be allotted when drilling solid without bushings. Record in miscellaneous column of machine time-rate sheet.
10. Use above feeds and speeds for reaming bored holes (.008-in. stock).
11. Use 2/3 above feeds for reaming drilled holes (1/64-in. stock).
12. When reaming, add .25 in. to length of travel for approach and overtravel.
13. Taper pin reamers may be run at drill feed and speed for largest diameter of taper.
14. When spot facing, gairing or fly-tool piloted, use counterbore speed and .003 feed.

data are illustrated in Figs. 2-7 through 2-10 and are discussed in greater detail in Chapter 6.

Handbook machinability ratings are helpful as starting points, but can never duplicate the unique conditions encountered in each production situation. Obviously, variables exist in tool materials, tool geometry, machine-tool condition, and workpiece material composition and properties. Therefore, adaptation and modification of published machinability data must be made no matter what the source.

FORMAT AND PROGRAM PREPARATION

The responsibility of manuscript preparation is normally charged to the parts programmer. The parts programmer may use a variety of programming sheets or coding forms upon which he writes the raw N/C processing data which is translated and converted, along with a suitable computer program and N/C machine language instructions, through a format into an input medium. In large N/C manufacturing

Table II-4. Speeds and Feed Times for Tapping.

Machinability Classification	N.C. and N.F. Taps								Pipe Taps		
	Size	Pitch	RPM	Time	Size	Pitch	RPM	Time	Size	RPM	Time
	20 sfpm (#10)				30 sfpm (#10)				20 sfpm		
	#8	32	611	.10	#8	36	611	.12	1/8	245	.19
	#10	24	611	.08	#10	32	611	.10	1/4	175	.23
	#12	24	611	.08	#12	28	611	.09	3/8	123	.29
	1/4	20	458	.09	1/4	28	458	.12	1/2	102	.33
3A	5/16	18	367	.10	5/16	24	367	.13	3/4	81	.40
and	3/8	16	306	.10	3/8	24	306	.16	1	64	.47
4A	7/16	14	262	.11	6/16	20	262	.15	1-1/4	51	.58
	1/2	13	229	.11	1/2	20	229	.17	1-1/2	44	.66
	9/16	12	204	.12	9/16	18	204	.18	2	35	.82
	5/8	11	184	.12	5/8	18	184	.20	2-1/2	28	1.04
	3/4	10	153	.13	3/4	16	153	.21	3	23	1.28
	7/8	9	131	.14	7/8	14	131	.21	3-1/2	20	1.50
	1	8	115	.14	1	14	115	.24	4	18	1.70
	1-1/8	7	102	.14	1-1/8	12	102	.24	4-1/2	16	1.95
	1-1/4	7	92	.15	1-1/4	12	92	.26	5	15	2.14
	1-1/2	6	76	.16	1-1/2	12	76	.32	6	13	2.56
	1-3/4	5	66	.15	1-3/4	12	66	.36			
	2	4-1/2	57	.16	2	12	57	.42			
	3	4	38	.21	3	10	38	.53			
	25 sfpm (#10)				40 sfpm (#10)				25 sfpm		
	#8	32	764	.08	#8	36	764	.09	1/8	306	.17
	#10	24	764	.06	#10	32	764	.08	1/4	218	.20
	#12	24	815	.06	#12	28	815	.07	3/8	153	.25
	1/4	20	611	.07	1/4	28	611	.09	1/2	127	.28
	5/16	18	489	.07	5/16	24	489	.10	3/4	101	.33
	3/8	16	408	.08	3/8	24	408	.12	1	81	.40
	7/16	14	349	.08	7/16	20	349	.11	1-1/4	64	.48
	1/2	13	306	.08	1/2	20	306	.13	1-1/2	55	.55
4B, 5B, 5C	9/16	12	272	.09	9/16	18	272	.13	2	43	.69
7A, 7B, 8A	5/8	11	245	.09	5/8	18	245	.15	2-1/2	35	.85
	3/4	10	203	.10	3/4	16	203	.16	3	29	1.04
	7/8	9	175	.10	7/8	14	175	.16	3-1/2	26	1.18
	1	8	153	.10	1	14	153	.18	4	23	1.35
	1-1/8	7	136	.10	1-1/8	12	136	.18	4-1/2	20	1.58
	1-1/4	7	123	.11	1-1/4	12	123	.20	5	18	1.80
	1-1/2	6	102	.12	1-1/2	12	102	.24	6	16	2.10
	1-3/4	5	87	.11	1-3/4	12	87	.28			
	2	4-1/2	76	.12	2	12	76	.32			
	3	4	51	.16	3	10	51	.39			

Notes: 1. Unit times are to feed in and out.
2. Time for pipe tapping includes depth control and reversal allowance of .10 min per hole.
3. For N.C. and N.F. taps add the following to tap depth: 1/8 in. for 18 pitch and over; 1/4 in. for 8–18 pitch; 1/2 in. for 8 pitch and under.
4. N.C. and N.F. taps—add .05 min for each hole for tap reversal with tap chuck or for through holes with tap and tap driver.
5. N.C. and N.F. taps—add .15 min to each hole for tap reversal and 1.0 in. to depth for blind holes tapped with tap and tap driver.
6. For holes tapped with N.C. or N.F. tap and tap driver, use 50 percent of speed given above. Unit time will then be 200 percent of above values.
7. To find the unit time per inch for feed in and out:

$$\text{Unit time per inch} = \frac{\text{Pitch}}{\text{rpm}} \times 2$$

RECOMMENDED MICROSTRUCTURES FOR ALLOY STEELS

Type Steel	Desirable Microstructure	Poor Machinability
Low Carbon 0.08–0.20	Ferrite plus pearlite Cold drawn	Abnormal Pearlite Coarse grain size
Medium Carbon 0.30–0.40	Lamellar Pearlite Uniform grain size	Segration, banding Free Ferrite
High Carbon 0.45–0.60	Spheroidized	Fine Pearlite

TYPICAL MICROSTRUCTURES IN ORDER OF DECREASING MACHINABILITY

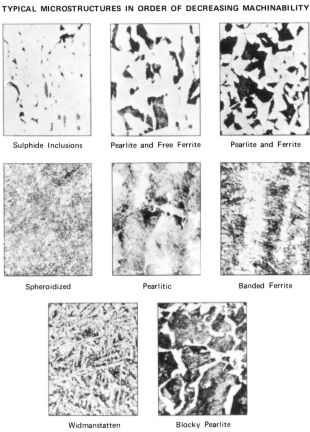

Sulphide Inclusions   Pearlite and Free Ferrite   Pearlite and Ferrite

Spheroidized   Pearlitic   Banded Ferrite

Widmanstatten   Blocky Pearlite

Fig. 2-7.—Effect of microstructure in carbon steels as related to machinability. *(Courtesy, Monarch Machine Tool Company, 10)*

Table II-5. Speeds and Feeds for

Machinability Classification	Depth of Cut					
	Fine Finish 1/16 in.		1/16. to 1/4 in.		1/4 in. to 1/2 in.	
	Speed	Feed	Speed	Feed	Speed	Feed
1A	60	1.1	50	2.7	50	2.2
2A	80	2.3	65	5.6	65	4.5
3A	100	2.8	80	7	80	5.6
4A	120	4	95	10	95	8
4B	120	4	95	10	85	6
5A	140	4.8	110	12	110	9.6
5B	140	4.8	110	12	110	9.6
5C	140	7.2	110	18	110	14
6A	160	6	125	15	125	12
7A	180	7.2	145	18	145	14
8A	200	10	160	25	160	20
9C	260	17	210	44	210	35
10D	450	24	350	57	350	46

Notes: 1. For plain milling cutters (without side teeth) use 80 percent of above feeds.
2. Speeds are given in sfpm. Use table to convert to rpm for proper cutting diameter.
3. For a single cut to size over 1/16 in. deep use 80 percent of above feeds.
4. Apply the following factors to feeds for varying rigidity: standard work conditions—1.0; very rigid conditions—1.5; semirigid conditions—.7; frail conditions—.4.
5. For staggered-tooth cutters apply factor of 1.5 to above feeds.
6. For carbide side mills use 2.5 times above speeds and 1.5 times above feeds.

facilities, several programmers may be involved in preparing manuscripts and machine control tapes. Each may be responsible for a designated group of N/C machines with varying degrees of capability and performance. Some organizations have such a commitment to N/C methods that large groups of men combine their talents to create the machine control tape. An effort of this type involves complex manuscript development and sophisticated computer programming. This is commonly the environment found in contour machining operations.

A manuscript prepared for the part shown in Fig. 2–11 and designed for the S/360 ADAPT/AUTOSPOT programming system is shown in Fig. 2–12. The input format is a symbolic-word, variable-length sentence structure that is recognizable by the ADAPT processor. Using this information, the processor, in conjunction with a post-processor, performs calculations to produce printed instructions as shown in Fig. 2–13. Further manipulation of these data and instructions results in a printout of machine control instructions as shown in Fig. 2–14.

The ADAPT processor will interpret any information in columns 1 through 72 of an input card. Card columns 73–80 may be used for card deck identification or sequencing. Card columns 1–80 will appear in the program listings shown in Figs. 2–13 and 2–14. Tool description, machining tolerance, and speed and feed data are used for program resolution (11).

Parts Programmer's Function

In a true sense, the parts programmer must possess knowledge and skill in three basic disciplines if he is to properly perform his duties. These disciplines are tool engi-

Side Milling and Plain Milling

Depth of Cut					
1/2 in. to 3/4 in.		3/4 in. to 1-1/16 in.		1-1/16 in. to 1-1/2 in.	
Speed	Feed	Speed	Feed	Speed	Feed
40	1.6	35	1.2	35	.7
55	3.4	50	2.5	50	1.4
70	4.2	60	3.2	60	1.8
85	6	70	4.5	70	2.5
85	6	70	4.5	70	2.5
100	7.2	85	5.4	85	3
100	7.2	85	5.4	85	3
100	11	85	8.1	85	4.5
110	9	95	6.8	95	3.8
125	11	110	8.1	110	4.5
140	15	120	11	120	6.3
180	26	150	20	150	11
325	34	270	26	270	14

neering, process engineering, and man-to-machine communications. The programmer must be able to use his knowledge and experience in these fields to effectively and efficiently control the motions and production output of an N/C machine tool.

The degree of knowledge and experience necessary largely depends upon the sophistication of the system. Two-axis machines, manually programmed, usually require a lesser degree of programming skill than multiaxis contouring machines using computer-aided methods. In either case, the programmer must know and be able to make use of tool and process engineering information. Knowledge of tools, cutting fluids, fixture design principles and techniques, and machinability data are important considerations in establishing machining control. Decisions founded upon this information will reveal themselves in the N/C machine control tape.

The knowledge required to activate the N/C programming system is just as important as tool and process information. Whether the system is simple or complex, the programmer must (1) analyze the part print, (2) decide on the sequence of operations, (3) write a manuscript compatible in language and format with the programming system which describes the geometric motion of the cutting tool and other functions, and (4) make use of the best and most reliable machinability data available. In addition he helps to design workholding devices and select tools, and he may supervise production scheduling.

Considerable work has been done to make the programming function less tedious and repetitive, less dependent upon analysis and calculations, and less subject to human error. This can be witnessed in the development of an Englishlike program vocabulary, the development of program subroutines, and the use of computers. The present-day

Table II-6. Speeds and Feeds for End-Mill

Diameter of End Mill (in.)	RPM		Depth of Cut							
	70 sfpm	85 sfpm	1/16		1/8		1/4		3/8	
			Chip Load	Feed	Chip Load	Feed	Chip Load	Feed	Chip Load	Feed
1/8	2139	2598	.0056	2.9	.00035	1.8	.00011	.6		
1/4	1070	1300	.0016	4.1	.00124	3.2	.0008	2.1	.0048	1.2
3/8	713	865	.0028	4.7	.0023	4.0	.0017	2.9	.0012	2.0
1/2	535	649	.0037	4.8	.0034	4.4	.0027	3.4	.0021	2.8
5/8	428	520	.0056	5.3	.0047	4.9	.0039	4.0	.0033	3.4
3/4	357	432	.0062	5.3	.0058	5.0	.0049	4.2	.0048	3.7
7/8	306	372	.0072	5.3	.0069	5.1	.0059	4.4	.0052	3.9
1	267	325	.0085	5.5	.0080	5.2	.0071	4.6	.0063	4.1
1-1/4	214	260	.0112	5.8	.0110	5.5	.0097	5.1	.0089	4.6
1-1/2	178	216	.0140	6.1	.0136	5.8	.0130	5.4	.0120	5.0
1-3/4	153	185	.0170	6.2	.0164	6.1	.0150	5.6	.0142	5.3
2	134	162	.0200	6.4	.0191	6.2	.0180	5.8	.0170	5.5

Notes: 1. Apply the following factors to feed rate for varying rigidity of workpiece or cutter: standard work conditions—1.0; very rigid conditions—1.5; semirigid conditions—.7; frail conditions—.4.

2. For milling sharp inside corners with sharp-cornered end mills use 75 percent of above feeds and speeds.

3. For boring into solids with end mills use 1-1/4-in. feed.

4. For milling other machinability classifications, apply the following factors to speeds and feeds for class 4B:

Machinability Classification	Chip Load Factor	RPM Factor	Feed Factor
1A	.67	.4	.27
2A	.84	.67	.5
3A	.84	.8	.7
4A	1	1	1
4B	1	1	1
5A	1	1.2	1.2
5B	1	1.2	1.2
5C	1.5	1.2	1.8
6A	1.2	1.3	1.5
7A	1.2	1.5	1.8
8A	1.5	1.7	2.5
9C	3	2.2	4.4
10D	1.5	3.8	5.7

programmer is required to analyze, select, and synthesize various kinds of data to produce an N/C machine control tape, but the future programmer may only have to activate, by switch or button, an N/C programming system comprising vast amounts of stored machinability data and computing programs.

STRINGENCY OF PROGRAMS

As discussed in Chapter 1, generalized machinability data usually provide only good starting points or general guidelines upon which effective machinability data can

Slotting—Machinability Classification 4B.

Depth of Cut											
1/2		5/8		3/4		1		1-1/4		1-1/2	
Chip Load	Feed	Chip Load	Feed	Chip Load	Feed	Chip Load	Feed	Chip Load	Feed	Chip Load	Feed
.00024	.6										
.00089	1.5	.00062	1.1	.00037	.6						
.0017	2.2	.00133	1.7	.00098	1.3	.00050	.6				
.0028	2.8	.0022	2.3	.0019	1.9	.00124	1.3	.00066	.7		
.0036	3.0	.0032	2.7	.0027	2.3	.00195	1.7	.00133	1.1	.0008	.7
.0047	3.5	.0042	3.1	.0036	2.7	.0027	2.0	.00204	1.5	.0014	1.1
.0057	3.7	.0051	3.3	.0045	2.9	.0035	2.3	.0028	1.8	.0021	1.4
.0080	4.1	.0074	3.8	.0067	3.5	.0057	3.0	.0046	2.4	.0039	2.0
.0110	4.6	.0097	4.2	.0092	4.0	.0079	3.4	.0069	3.0	.0058	2.5
.0131	4.9	.0123	4.6	.0116	4.3	.0102	3.8	.0092	3.3	.0080	3.0
.0160	5.2	.0150	4.8	.0142	4.6	.0128	4.1	.0114	3.7	.0102	3.3

be based. Usually, the parts programmer has only this type of information at his dis-posal—information which can, at best, only approximate the actual machining condi-tions in the production environment which usually exists for either manual or computer-aided programming methods.

The ideal method of production would be to apply no further changes and no shop overrides once the N/C control tape has been created. Theoretically, changes in the system after the machine control tape has been generated defeat the purpose of the N/C concept; that is, interference with the control by process changes restricts full implementation of the ultimate control concept. In practice, however, changes in the application of input data by speed and, particularly, feed overrides on the control system must often be made.

Machining Disturbances

The parts programmer is responsible for entering information by format and pro-gram to prepare a control tape that reflects the best machinability data available for an effective operation sequence. He is also responsible for revising that information. Any changes which must be made to this input data result from a lack of control over metal-cutting variables—disturbances in the machine-tool operation. These variables are listed as follows:

1) Workpiece Variations—Workpiece characteristics of size and shape, chemical composition, and physical and mechanical properties may appear which differ from those originally predicted.

2) Setup Errors—Errors in the initial setup points used to begin the N/C opera-tion are often uncovered by proving the tape. These errors must be corrected to obtain a correct relationship between the tool home position, setup points, and the location of the workpiece mounted in its fixture.

Table II-7. Speeds and Feeds for End-Milling Using Side of

Diameter of End Mill (in.)	RPM		Depth of Cut							
	70 sfpm	85 sfpm	1/16		1/8		3/16		1/4	
			Chip Load	Feed	Chip Load	Feed	Chip Load	Feed	Chip Load	Feed
1/8	2139	2598	.00035	1.8						
1/4	1070	1300	.00124	3.2	.0008	2.1				
3/8	713	865	.0023	4.0	.0017	2.9	.00124	2.1		
1/2	535	649	.0034	4.4	.0027	3.4	.0021	2.8	.0017	2.2
5/8	428	520	.0047	4.9	.0034	4.0	.0033	3.4	.0028	2.8
3/4	357	432	.0058	5.0	.0049	4.2	.0043	3.7	.0036	3.1
7/8	306	372	.0069	5.1	.0059	4.4	.0052	3.9	.0047	3.5
1	267	325	.0080	5.2	.0071	4.6	.0063	4.1	.0057	3.7
1-1/4	214	260	.0110	5.5	.0097	5.1	.0089	4.6	.0080	4.1
1-1/2	178	216	.0136	5.8	.0130	5.4	.0120	5.0	.0110	4.6
1-3/4	153	185	.0164	6.1	.0150	5.6	.0142	5.3	.0131	4.9
2	134	162	.0190	6.2	.0180	5.8	.0170	5.5	.0160	5.2

Notes: 1. Use the feeds given above when the width of cut is approximately equal to the diameter of the end mill.

2. For widths of cut not equal to the end-mill diameter, apply the following factors to above feeds:

Width of Cut Along Axis	Feed Factor
1/4 cutter diameter	2
1/3 cutter diameter	1.85
1/2 cutter diameter	1.6
3/4 cutter diameter	1.25
1-1/4 cutter diameters	.8
1-1/2 cutter diameters	.6
1-3/4 cutter diameters	.45
2 cutter diameters	.3

3. For milling other machinability classifications, apply the following factors to speeds and feeds for class 4B:

Machinability Classification	Chip Load Factor	RPM Factor	Feed Factor
1A	.67	.4	.27
2A	.84	.67	.5
3A	.84	.8	.7
4A	1	1	1
4B	1	1	1
5A	1	1.2	1.2
5B	1	1.2	1.2
5C	1.5	1.2	1.8
6A	1.2	1.3	1.5
7A	1.2	1.5	1.8
8A	1.5	1.7	2.5
9C	3	2.2	4.4
10D	1.5	3.8	5.7

4. Apply the following factors to feed rate for varying rigidity of workpiece or cutter: standard work conditions—1.0; very rigid conditions—1.5; semirigid conditions—.7; frail conditions—.4.

5. For milling sharp inside corners with sharp-cornered end mills use 75 percent of above feeds and speeds.

Cutter—Machinability Classification 4B.

Depth of Cut											
3/8		1/2		5/8		3/4		7/8		1	
Chip Load	Feed	Chip Load	Feed	Chip Load	Feed	Chip Load	Feed	Chip Load	Feed	Chip Load	Feed
.0027	2.3										
.0036	2.7										
.0045	2.9	.0035	2.3								
.0067	3.5	.0057	3.0	.0046	2.4						
.0092	4.0	.0079	3.4	.0069	3.0	.0058	2.5				
.0116	4.3	.0102	3.8	.0092	3.3	.0080	3.0	.0068	2.5		
.0142	4.6	.0128	4.1	.0114	3.7	.0102	3.3	.0089	2.9	.0080	2.6

6. When using multiple-lip end mills, increase feed the following amounts:

Number of Lips	Feed Increase
4	20 percent
6	35 percent
8	45 percent

3) Motion Errors—Motion errors are a common source of control tape changes. A cutting tool moving in its six degrees of freedom may interfere with work locating devices and clamping systems as well as produce machining errors in size, shape, and finish.

4) Speed/Tool-Life Errors—The speed/tool-life relationship may be inaccurately programmed. It has been established that speed, feed, type of material, and tool geometry are the major factors affecting tool life. It may be necessary for the programmer to revise these factors in order to achieve predictable tool life.

5) Surface Finish Variations—Feed rate, cutting speed, tool material, tool geometry, cutting fluid, and workpiece material are variables that are important in achieving a controlled finish. These variables may need to be revised to obtain a correct master control tape.

Tape Testing

After the control tape is loaded into the N/C machine control system and the workpiece and tools are installed, usual shop procedure is to "jog" the N/C machine through its various control attitudes as the tape directs. It is at this point that the programmer can begin to see incorrectly programmed information. However, jogging the machine tool through its motions does not always give a clear or accurate picture of how the cutting tool will respond to the workpiece material. No information is gained about tool life, surface finish, or horsepower requirements, for example. Also, the parts programmer has no way of making a critical evaluation of the machinability data he used in producing the tape.

Table II-8. Speeds and Feeds for Face Milling with T.C. Full Back Cutters.

Machinability Classification	Type Cut	Depth of Cut (in.)	Speed (sfpm)	Feed	Chip Load per Tooth	Cutter (dia.–no. teeth)								
						3–8 RPM	4–10 RPM	5–10 RPM	6–12 RPM	7–16 RPM	8–18 RPM	9–20 RPM	10–22 RPM	12–26 RPM
4B Hard Cast Iron	Fine	up to 1/16	300	10	.004	382	286	230	190	165	143	127	115	95
	Standard	up to 1/16	270	20	.008	344	260	205	170	147	130	115	103	85
	Rough	1/16 to 3/16	240	16	.008	305	230	183	153	130	115	102	93	75
		1/4 to 1/2	240	13	.008	305	230	183	153	130	115	102	93	75
5B Medium Cast Iron	Fine	up to 1/16	350	13	.004	445	335	267	223	190	167	150	135	110
	Standard	up to 1/16	325	25	.008	414	310	248	206	177	155	138	124	103
	Rough	1/16 to 3/16	280	20	.008	357	267	215	178	153	134	120	107	90
		1/4 to 1/2	280	16	.008	357	267	125	178	153	134	120	107	90
5C Hard Bronze	Fine	up to 1/16	350	13	.0045	455	335	267	223	190	167	150	135	110
	Standard	up to 1/16	325	25	.009	414	310	248	206	177	155	138	124	103
	Rough	1/16 to 3/16	280	20	.009	357	267	215	178	153	134	120	107	90
		1/4 to 1/2	280	20	.009	357	267	215	178	153	134	120	107	90
9C Soft Bronze	Fine	up to 1/16	625	25	.005	796	597	478	399	341	292	265	238	199
	Standard	up to 1/16	575	50	.010	732	549	439	366	314	274	244	220	183
	Rough	1/16 to 3/16	525	40	.010	670	500	400	330	290	250	220	200	167
		1/4 to 1/2	525	35	.010	670	500	400	330	290	250	220	200	167
10B Aluminum	Fine	up to 1/16	1100	35	.004	1400	1050	840	699	600	524	467	420	349
	Standard	up to 1/16	1000	65	.008	1273	955	764	637	546	477	424	382	318
	Rough	1/16 to 3/16	900	55	.008	1146	859	688	573	491	430	382	344	286
		1/4 to 1/2	900	48	.008	1146	859	688	573	491	430	382	344	286

Notes: 1. If rpm available on the machine is lower than the above recommendations, feed must be reduced to maintain finish.

2. For single cut to size over 1/16 in. deep use 80 percent rough cut feed.

3. Apply the following factors to feed for varying rigidity of cutter or workpiece: standard work conditions—1.0; very rigid conditions—1.5; semirigid conditions—.7; frail conditions—.4.

4. Fine finish recommendations to be used only when surfaces must be very flat and smooth.

5. For finish for good appearance on parts that will show, or for faces that are to be buffed, use feed midway between fine and standard finish feeds and use fine finish rpm.

6. For HSS face mills on rough cuts use 40 percent rpm and 60 percent feed. On finish cuts use 40 percent rpm and 40 percent feed.

MACHINABILITY RATINGS

TYPE	COMPOSITION					BRINELL	TENSILE PSI	MACHIN-ABILITY INDEX
	TC	SI	MN	CR	NI			
Gray Iron								
Class 30 Ferritic	3.40	2.10	0.60			140-179	30,000	82
Class 40 Pearlitic	3.20	1.60	0.90			207-229	45,000	68
Nodular								
60-45-10 Ferritic	3.52	2.56	0.40		1.13	140-200	70,700	110
80-60-03 Pearlitic	3.40	2.82	0.42		1.00	200-270	87,900	80
100-70-03 H.T.						270-300	105,000	65
Meehanite GA	3.12	1.17	0.96			196-217	50,000	85
Malleable								
53004-(85M)	2.50	1.05	0.96			197-294	80,000	75
35018-(86M)	2.40	1.17	0.50			110-156	35,000	90
N. Resist	3.10	2.10	0.80		16.0	140-170		45
Chilled (White)	3.50	0.75	0.60		5.50	470-650		15

TYPICAL MICROSTRUCTURES IN ORDER OF DECREASING MACHINABILITY

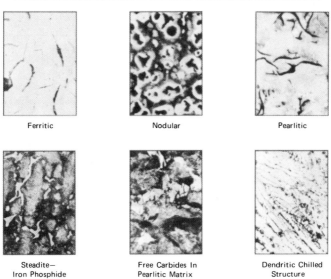

Ferritic	Nodular	Pearlitic
Steadite— Iron Phosphide	Free Carbides In Pearlitic Matrix	Dendritic Chilled Structure

Fig. 2–8.—Machinability of cast iron directly related to physical properties, micro-structure, and chemical composition. *(Courtesy, Monarch Machine Tool Company, 10)*

Table II-9. Speeds and Feeds for Open Boring

Bore Diameter (in.)	Depth of Cut—Radius (in.)	Carbide Tools			
		Speed (sfpm)	RPM	Boring Bars	
				Feed	Unit
Through 5/8	1/32		(Use HSS Tools)		
	.010		(Use HSS Tools)		
5/8–1	1/16	125	478	.004	.53
	.010	100	382	.003	.87
1–1 1/4	1/16	150	459	.004	.55
	.015	110	337	.003	1.00
1 1/4–1 1/2	1/8	150	381	.006	.44
	1/16	185	470	.004	.53
	.015	140	356	.003	.94
1 1/2–1 3/4	1/8	150	327	.006	.51
	1/16	185	403	.004	.62
	.015	140	305	.003	1.09
1 3/4–2	3/16	155	296	.006	.56
	1/8	155	296	.008	.42
	1/16	200	382	.005	.52
	.015	170	325	.003	1.03
2–2 1/4	3/16	160	272	.008	.46
	1/8	160	272	.010	.37
	1/16	235	398	.008	.31
	.015	195	326	.003	1.02
2 1/4–2 1/2	1/4	160	245	.008	.51
	1/8	160	245	.019	.41
	1/16	240	367	.008	.34
	.015	200	306	.003	1.09
2 1/2–2 3/4	3/8	165	229	.008	.55
	1/8	165	229	.010	.44
	1/16	240	333	.008	.38
	.015	205	285	.003	1.17
2 3/4–3	3/8	170	216	.010	.46
	1/16	245	312	.008	.40
	.015	210	267	.003	1.25
3–3 1/4	3/8	170	199	.010	.50
	1/16	245	288	.010	.35
	.015	230	270	.003	1.23
3 1/4–3 1/2	3/8	175	191	.012	.44
	1/16	245	268	.010	.36
	.015	250	272	.003	1.23
3 1/2–4	3/8	180	172	.016	.36
	1/16	245	234	.012	.36
	.015	270	258	.003	1.29
4–4 1/2	3/8	185	157	.016	.40
	1/16	245	208	.012	.40
	.015	270	229	.003	1.46
4 1/2–16	3/8	200		.018	
	1/16	245		.012	
	.015	290		.003	

Notes: 1. Approach and overtravel in the clear = .07 in. Approach and overtravel inside = .13 in.
2. Do not exceed recommended values by more than 10 percent to attain speed available on the machine. Use closest feed available.

Carbide Tools		Speed (sfpm)	HSS Tools		
Duplex Bars			RPM	Feed	Unit
Feed	Unit				
		50	306	.005	.65
		50	306	.005	.65
.003	.70	50	191	.005	1.05
.003	.87	50	191	.005	1.05
.003	.73	50	153	.006	1.09
.003	1.00	50	153	.004	1.31
.006	.44	50	127	.007	1.12
.006	.35	60	153	.006	1.09
.003	.94	50	127	.005	1.57
.006	.51	50	109	.007	1.31
.006	.41	60	131	.006	1.27
.003	1.09	50	109	.005	1.84
.006	.56	50	96	.007	1.49
.006	.56	50	96	.009	1.16
.006	.44	65	124	.007	1.15
.003	1.03	55	105	.005	1.91
.006	.61	50	86	.009	1.29
.006	.61	50	86	.011	1.05
.006	.43	70	119	.009	.93
.003	1.02	60	102	.006	1.64
.006	.68	55	84	.009	1.32
.006	.68	55	84	.011	1.08
.006	.45	75	114	.009	.97
.003	1.09	65	100	.006	1.67
.006	.73	55	77	.012	1.08
.006	.73	55	77	.012	1.08
.006	.50	75	104	.010	.96
.003	1.17	65	90	.006	1.85
.006	.77	60	76	.016	.82
.006	.53	80	102	.010	.98
.003	1.25	75	95	.006	1.76
.006	.84	60	70	.016	.89
.006	.58	80	94	.010	1.06
.003	1.23	80	94	.006	1.78
.006	.87	60	66	.016	.95
.006	.62	80	87	.010	1.15
.003	1.23	85	93	.006	1.80
.006	.97	60	57	.019	.92
.006	.71	80	76	.012	1.10
.003	1.29	85	81	.006	2.06
.006	1.06	60	51	.019	1.03
.006	.80	80	68	.012	1.22
.003	1.46	90	72	.006	2.32
.006		65		.019	
.006		80		.012	
.003		90		.008	

3. *Boring bars only*—do not apply factors to feeds for .010-in. or .015-in. depths of cut. Do not apply factors to other feeds below the feeds for .010-in. or .015-in. depths of cut.

Table II-10. IBM Division Standard—Machinability Classifications.

IBM Code	Material	Machinability Classification
03-610	Sintered Brass, 70-30	
03-613	Brass, Sintered, Density 7.0 to 7.5 g/cc	Free
03-618	Brass, Sintered, Density 7.5 g/cc Min	Free
03-650	Bronze Phosphor ASTM B 144 Alloy 3A	
03-653	Casting Alloy	Free
03-700	Bronze, ASTM B138, Alloy B	
03-702	ASTM B138, Alloy B, Half Hard Temper	Medium
03-703	ASTM B138, Alloy B, Hard Temper	Medium
03-710	Bronze, Phosphor, ASTM B139, Alloy D	
03-712	ASTM B139, Alloy D, Cold Drawn	Hard
03-720	Bronze Tin ASTM B 143 Alloy 1A	
03-721	Casting Alloy	Hard
03-730	Bronze Tin, ASTM B143, Alloy 2A	
03-731	ASTM B143, Alloy 2A, Navy M Sand Cstg	Medium
03-740	Bronze Aluminum ASTM B148 Alloy 9A	
03-741	Casting Alloy	Hard
03-750	Bronze Silicon	
03-751	Casting Alloy	Hard
03-800	Nickel Silver, ASTM B151, Alloy A	
03-802	ASTM B151, Alloy A, Hard Temper	Hard*
03-820	Nickel Silver, ASTM B151, Alloy C	
03-825	ASTM B151, Alloy C, Quarter Hard Temper	Hard*
03-830	Sintered Bronze, Structural	
03-835	7.0 g/cc Minimum Density	Medium Low
03-840	Sintered Bronze, Bearing	
03-845	Density 6.4 to 6.8 g/cc	Medium Low
03-850	Sintered Bronze, Leaded Bearing	
03-855	Density 6.5 to 6.9 g/cc	Medium Low
03-860	Bronze Phosphor ASTM B103 Alloy C	
03-861	Hard	Hard
03-870	Bronze, Phosphor, ASTM B139, Alloy A	
03-871	ASTM B139, Alloy A, Hard Drawn Temper	Hard
03-910	Bronze, Lead-Phos. ASTM B139, Alloy B1	
03-911	ASTM B139, Alloy B1, Cold Rolled	Hard
03-912	ASTM B139, Alloy B1, Cold Drawn	Hard
03-920	Bronze, Lead-Phos. ASTM B139, Alloy B2	
03-923	ASTM B139, Alloy B2, Hard Temper	Free
03-950	Nickel-Silver, ASTM B122, Alloy 3	
03-951	ASTM B122, Alloy 3, Hard Temper	Hard*
03-952	ASTM B122, Alloy 3, Half Hard Temper	Hard*
03-955	ASTM B122, Alloy 3, Quarter Hard Temper	Hard*
03-970	Copper-Silicon, ASTM B97, Alloy A	
03-971	ASTM B97, Alloy A, Annealed Temper	Medium
03-972	ASTM B97, Alloy A, Quarter Hard Temper	Medium
03-973	ASTM B97, Alloy A, Half Hard Temper	Medium
03-974	ASTM B97, Alloy A, Hard Temper	Medium
03-975	ASTM B97, Alloy A, Extra Hard Temper	Medium
03-976	ASTM B97, Alloy A, Spring Temper	Medium
03-990	Nickel-Silver, ASTM B122, Alloy 8	
03-991	ASTM B122, Alloy 8, Spring Temper	Hard
03-993	ASTM B122, Alloy 8, Hard Temper	Hard
03-994	ASTM B122, Alloy 8, Extra Hard Temper	Hard
03-995	ASTM B122, Alloy 8, Half Hard Temper	Hard

*Use bronze, brass, and copper.
†Use malleable and cast iron.

Table II-10. IBM Division Standard—Machinability Classifications. (*Continued*)

IBM Code	Material	Machinability Classification
05-090	Iron Wrought Products	
05-091	Wrought Iron	Good
05-150	Cast Iron	
05-151	Iron Castings, Gray	Good
05-152	Iron, Gray, Nickel Bearing	Good
05-153	Iron, Malleable Castings—Ferritic	Excellent
05-154	Iron, Malleable—Pearlitic	Excellent
05-155	Iron, Gray, High Strength	Fair
05-156	Iron, Ductile Castings—Ferritic	Excellent
05-157	Iron, Ductile Castings—Pearlitic	Excellent
05-158	Iron, Cast Alloy, Meehanite	Good
05-159	Miscellaneous Cast Iron	Good
05-190	Sintered Iron	
05-191	Iron, Sintered, 6.0 g/cc Min Density	Medium Low
05-192	Iron, Sintered, 7.0 g/cc Min Density	Medium Low
05-193	Iron, Sintered, 7.3 g/cc Min Density	Medium Low
05-194	Iron, Sintered, 7.3 g/cc Min Density.	Medium Low
05-195	Iron, Sintered, 7.5 g/cc Min Density	Medium Low
05-196	Iron, Sintered, 7.5 g/cc Min Density.	Medium Low
05-200	Sintered Iron-Copper	
05-204	Iron Copper, Sintered, High Copper	Hard
05-205	Iron Copper, Sintered, Low Copper	Hard
05-206	Iron Copper, Sintered, Low Density	Hard
05-208	Iron Copper, Sintered, Intermediate Den	Hard
05-210	Sintered Iron, Electrical	
15-215	Iron, Sintered, Electrical	Hard
05-770	Magnet Steels	
05-771	3 5 Pc Chromium ·	Fair†
05-772	6 0 Pc Tungsten	Fair†
05-820	Sintered Iron, Cu or Brs Infiltrated	
05-825	Iron, Sintered, Copper	Hard
05-870	Iron, Silicon 4.0%	Fair
05-872	Cold Rolled	Fair
05-880	Iron, Silicon 1.0%	
05-881	Silicon Iron, 1.0% Cold Drawn	Fair†
05-882	Silicon Iron, 1.0% Cold Rolled	Fair†
05-890	Iron, Silicon 2.5%	
05-891	Silicon Iron 2.5% Cold Drawn	Fair
05-892	Silicon Iron 2.5% Cold Rolled	Fair
05-900	Iron, Electrical	
05-901	Electrical Iron, Intermediate Temper	Fair†
05-902	Electrical Iron, Cold Drawn	Fair†
05-904	Electrical Iron, Annealed Temper	Fair†
05-905	Electrical Iron, Typewheels	Fair†
05-906	CR RB 65 Max	Fair†
05-907	Forging Quality	Fair†
05-930	Iron-Nickel Alloy Pendulum Material	
05-932	Invar 36	Fair†
05-940	Iron-Nickel Alloy High Nickel	
05-942	Carpenter Hy Mu 80	Fair†
05-943	Mu Metal	Fair†
05-950	Iron-Nickel Alloy Medium Nickel	
05-951	Iron-Nickel Alloy, Cold Rolled	Fair
05-952	Iron-Nickel Alloy, Cold Drawn	Fair

*Use bronze, brass, and copper.
†Use malleable and cast iron.

Table II-10. IBM Division Standard—Machinability Classifications. (*Continued*)

IBM Code	Material	Machinability Classification
06-050	Steel, Carbon, Vinyl Covered	
06-054	Vinyl Covered Steel	Medium Low
06-055	Vinyl Covered Steel, Stretcher Leveled	Medium Low
06-056	Vinyl Covered Steel, Deep Drawing	Medium Low
06-060	Steel, Carbon, AISI C-1010	
06-061	AISI C-1010, No 1 Temper	Medium High
06-062	AISI C-1010, No 2 Temper	Medium High
06-063	AISI C-1010, No 3 Temper	Medium Low
06-064	AISI C-1010, No 4 Temper	Medium Low
06-065	AISI C-1010, No 5 Temper	Medium Low
06-066	AISI C-1010, Card Racks	Medium Low
06-067	AISI C-1010, Cold Rolled	Medium Low
06-069	Special Shapes	Medium Low
06-070	Steel, Carbon, AISI C-1010	
06-071	AISI C-1010, Aluminum Killed	Medium High
06-076	AISI C-1008 or C-1010 Card Rack	Medium Low
06-079	AISI C-1010 Special Section	Medium Low
06-110	Steel, Carbon, AISI C-1015	
06-111	AISI C-1015, Hot Rolled	Medium Low
06-120	Steel, Carbon, AISI MT-1015	
06-121	AISI MT-1015, Finish Annealed Temper	Medium High
06-126	AISI MT-1015, Annealed Temper.	Medium Low
06-190	Carbon Steel Low Carbon Gen Purpose	
06-191	Cold Finish	Medium High
06-192	Hot Finish	Medium Low
06-200	Steel, Carbon, AISI C-1020	
06-201	AISI C-1020, Aluminum Killed	Medium High
06-203	AISI C-1020, Rb 70-85	Medium High
06-207	AISI C-1020, Forging Quality	Medium Low
06-208	AISI C-1020, Investment Casting	Medium High
06-209	AISI C-1020, Rb 60-70	Medium Low
06-210	Steel, Carbon, AISI C-1025	
06-211	AISI C-1025, Cold Drawn	Medium High
06-212	AISI C-1025, Casting	Medium Low
06-213	AISI C-1025, Modified Tubing	Medium High
06-220	Steel, Carbon, AISI C-1018	
06-221	AISI C-1018, Cold Finished	Medium High
06-223	AISI C-1018, Hot Rolled	Medium Low
06-227	AISI C-1018, Forging Quality	Medium Low
06-228	AISI C-1018, Ground & Polished	Medium High
06-230	Steel, Carbon, Leaded AISI C-1018	
06-231	AISI C-1018, Leaded, Cold Rolled	Free
06-350	Steel, Carbon, AISI C-1035	
06-352	AISI C-1035, Rb 90-105	Medium High
06-353	AISI C-1035, Rb 80-90	Medium High
06-354	AISI C-1035, Rb 70-80	Medium Low
06-355	AISI C-1035, Rb 60-70	Medium Low
06-400	Steel, Carbon, AISI C-1040	
06-401	AISI C-1040, Cold Finished	Medium High
06-440	Carbon Steel AISI C 1042 Analysis	
06-441	CD Steel	Medium High

Table II-10. IBM Division Standard—Machinability Classifications. *(Continued)*

IBM Code	Material	Machinability Classification
06-480	Steel, Carbon, AISI C-1050	
06-482	AISI C-1050, Rb 90-105	Medium High
06-483	AISI C-1050, Rb 80-90	Medium High
06-484	AISI C-1050, Pretempered	Medium Low
06-485	AISI C-1050, Rb 70-85	Medium Low
06-530	Steel, Carbon, AISI C-1065	
06-531	AISI C-1065, Cold Drawn	Medium Low
06-570	Steel, Carbon, AISI C-1070	
06-572	AISI C-1070, Rb 85-95	Medium Low

Table II-11. IBM Division Standard—Speeds and Feeds for Plain Drilling.

Material	Drill Diameter	Machinability Classification			
		Superior and Excellent (225 sfpm)		Good and Fair (190 sfpm)	
		RPM	Feed (in./min)	RPM	Feed (in./min)
Aluminum	.040		5.70		4.80
and	.062		8.30		7.00
Magnesium	.125	6875	14.90	5805	12.80
	.188	4583	17.00	3870	14.30
	.250	3437	19.00	2902	16.00
	.312	2689	16.00	2322	13.70
	.375	2240	13.00	1935	11.20
	.438	1964	12.50	1658	10.60
	.500	1718	12.00	1451	10.00
	.562	1527	11.00	1290	9.40
	.625	1375	10.00	1161	8.10
	.688	1250	9.00	1055	7.50
	.750	1145	7.80	967	6.90
	.812	1057	7.10	893	6.30
	.875	982	6.40	829	5.40
	.938	915	5.70	772	4.80
	1.000	859	5.00	725	4.20
	1.062	806		681	
	1.125	763		645	
	1.188	723		611	
	1.250	687		580	

Table II-11. IBM Division Standard—Speeds and Feeds for Plain Drilling.
(Continued)

	Drill Diameter	Free 390 sfpm		Medium 180 sfpm		Hard 190 sfpm	
		RPM	Feed (in./min)	RPM	Feed (in./min)	RPM	Feed (in./min)
Copper,	.040		11.70		6.10		4.00
Brass,	.062		19.20		11.50		6.00
and Bronze	.125		36.90	5500	22.40	3055	10.50
	.188		23.00	3666	18.10	2037	8.45
	.250		21.20	2750	14.37	1527	7.13
	.312	4277	20.20	2200	12.20	1222	6.15
	.375	3566	19.00	1833	9.80	1018	4.92
	.438	3057	18.00	1571	8.90	873	4.50
	.500	2675	17.00	1375	8.30	763	3.98
	.562	2377	15.20	1222	6.70	679	3.00
	.625	2139	13.50	1100	6.00	611	2.98
	.688	1941	11.60	1000	5.80	555	2.50
	.750	1783	9.60	916	4.80	509	2.05
	.812	1646	8.20	846	4.10	470	1.90
	.875	1528	7.00	785	3.50	436	1.60
	.938	1426	5.80	735	2.90	408	1.38
	1.000	1337	4.80	687	2.40	381	1.23
	1.062	1259		647		358	
	1.125	1188		611		339	
	1.188	1069		479		322	
	1.250	891		550		305	

	Drill Diameter	Superior 210 sfpm		Excellent 150 sfpm		Free 125 sfpm		Med. High 90 sfpm		Med. Low 65 sfpm		Hard 50 sfpm	
		RPM	Feed (in./min)	RPM	Feed (in./min)	RPM	Feed (in./min)	RPM	Feed (in./min)	RPM	Feed (in./min)	RPM	Feed (in./min)
Carbon	.040		6.75		5.30		4.80		4.00		3.00		2.50
and	.062		10.30		9.30	7697	9.00	5541	6.00	4012	5.40	3057	3.75
Alloy	.125	6417	19.50	4583	19.50	3820	16.40	2750	10.50	1986	9.25	1528	6.40
Steels	.188	4278	21.00	3055	15.75	2548	14.20	1833	9.40	1324	7.50	1019	5.30
	.250	3208	22.80	2291	12.50	1910	12.30	1375	8.50	993	5.75	764	4.30
	.312	2566	17.50	1833	10.60	1527	9.90	1100	7.50	794	4.80	611	3.00
	.375	2139	11.75	1527	8.50	1273	7.75	916	6.25	662	3.60	509	1.70
	.438	1833	11.30	1309	7.80	1093	6.90	785	5.60	567	3.40	437	1.40
	.500	1604	10.80	1145	7.25	955	6.20	687	4.75	496	3.20	382	1.20
	.562	1426	10.20	1018	5.80	869	5.00	611	4.30	441	1.70	340	1.15
	.625	1283	9.40	916	4.60	765	4.10	550	3.75	397	2.20	306	1.10
	.688	1166	7.30	833	3.70	692	3.50	500	3.30	361	1.70	273	1.05
	.750	1069	5.20	763	2.75	635	2.70	458	2.60	331	1.50	254	1.00
	.812	987	4.60	705	2.70	593	2.50	423	2.40	305	1.40	237	.70
	.875	916	4.20	654	2.40	548	2.10	392	1.90	283	1.30	219	.50
	.938	855	3.70	610	2.30	509	1.90	366	1.50	265	1.25	204	.30
	1.000	802	3.00	572	1.90	478	1.55	343	1.25	248	1.20	191	.25
	1.062					449		324		234		180	
	1.125					425		305		220		170	
	1.188							289		209		161	
	1.250							275		198		153	

Notes:
1. Drill relief time must be provided if hole depth is greater than three (3) times drill diameter.
2. Feed in in./min can be maintained, on a given drill and material, with considerable variance from the recommended rpm, which is nonmandatory.
3. Drilling speeds are theoretical for the sfpm used.
4. Actual feed rates are in./min. When long series drills are required, use: .5 times

Table II-11. IBM Division Standard—Speeds and Feeds for Plain Drilling.
(Continued)

	Drill Diameter	Excellent 130 sfpm		Good 80 sfpm		Fair 60 sfpm	
		RPM	Feed (in./min)	RPM	Feed (in./min)	RPM	Feed (in./min)
Malleable and Cast Iron	.040		8.00		6.59		3.50
	.062		10.00	4871	10.50	3750	4.00
	.125	3973	20.00	2445	12.60	1833	6.90
	.188	2649	16.80	1630	11.00	1222	5.50
	.250	1986	13.70	1222	9.00	916	4.50
	.312	1589	12.70	978	8.20	733	3.20
	.375	1323	11.70	815	7.30	611	2.50
	.438	1136	10.20	699	6.50	523	2.20
	.500	993	8.70	611	5.70	458	2.00
	.562	883	7.70	543	5.30	401	1.80
	.625	796	6.70	489	4.70	366	1.50
	.688	722	5.60	444	4.50	333	1.30
	.750	661	4.40	409	4.20	305	1.00
	.812	616	3.90	379	3.60	282	.70
	.875	569	3.30	349	3.00	261	.50
	.938	529	2.80	326	2.50	244	.30
	1.000	497	2.30	306	2.20	229	.25
	1.062	467		287		216	
	1.125	442		272		203	
	1.188	419		258		193	
	1.250	398		245		185	

	Drill Diameter	Free 90 sfpm		Medium 70 sfpm		Hard 45 sfpm	
		RPM	Feed (in./min)	RPM	Feed (in./min)	RPM	Feed (in./min)
Stainless Steel	.040		5.20		3.70		2.30
	.062	5502	9.20	4280	6.50	2751	3.90
	.125	2750	13.00	2139	10.80	1375	8.50
	.188	1834	12.50	1426	9.10	917	5.70
	.250	1376	11.70	1070	7.50	688	3.40
	.312	1100	8.60	856	5.95	550	3.35
	.375	916	5.60	713	4.40	458	3.30
	.438	786	4.70	611	3.70	393	2.70
	.500	688	3.90	535	3.20	344	2.40
	.562	611	3.00	475	2.50	306	2.00
	.625	552	2.20	428	2.00	276	1.80
	.688	500	2.00	389	1.70	249	1.50
	.750	458	1.80	357	1.60	229	1.30
	.812	427	1.60	332	1.30	213	1.00
	.875	392	1.50	306	1.10	196	.70
	.938	366	1.30	285	.90	183	.50
	1.000	344	1.20	267	.80	172	.30
	1.062	323		251		162	
	1.125	306		238		153	
	1.188	290		225		145	
	1.250	274		214		137	

feed for .040-in. to .090-in. drill diameters; .7 times feed for .090-in. to .187-in. drill diameters; .8 times feed for .187-in. to .250-in. drill diameters.

5. For flat-bottom drills, counterbores, countersinks, and core drills use .7 times theoretical drilling rpm for speed and actual recommended feed.

6. For dwell, add .032 in. to length of cut for these types of tools with tolerance on depth over .002 in. Add .047 to length of cut for these types of tools with tolerance on depth up to and including .002 in.

RESULFURIZED STEELS

	BHN	Index
B-1112 CD	163	100
B-1113 CD	170	132
RSC-12L13 (Leaded)	140	160
C-1117 CD	170	87
C-11L17 (Leaded)	170	106
C-1137 HR	187	64
C-1137 CD	197	71
C-1141 HR	207	60
C-1141 CD	240	62
C-1144 HR	207	66
C-1145 HR	217	55

PLAIN CARBON STEELS

	BHN	Index
C-1008 (Rimmed)	97	64
C-1008 CD	107	71
C-1018 CD	163	74
C-1020 HR	143	76
C-1020 CD	156	80
C-1025 HR	143	61
C-1025 CD	163	67
C-1035 HR	179	59
C-1035 CD	201	64
C-10L45 (Leaded)	197	67
C-1045 HR	217	50
C-1045 CD	256	52
C-1050 HR	223	45
C-1095 A	243	60

CARBURIZING STEELS

	BHN	Index
3115 HR	160	58
E-3310 A	196	41
4027 HR	190	70
4023 CD	183	72
4320 A	174	52
4615 HR	183	58
4615 CD	217	64
4815 HR	196	45
6120 CD	187	52
8617 HR	197	70
8617 CD	212	80
8620	197	73
E-9310 CD	223	41

MEDIUM CARBON STEELS

	BHN	Index
1330-N	223	60
2330-A	192	50
2330-CD	212	45
3140 HR	197	57
3140 CD	229	64
4042 HR	235	67
4130 HR	183	65
4130 CD	201	70
4140 HR	187	60
4145 HR	217	55
4340 HR	223	52
4340 CD	248	47
4640 HR	187	55
5045	197	60
5145	217	52
E-6150 HR	197	52
E-6180 A	207	43
8640 HR	183	60
E-8740 HR	217	57
E-52100	192	54
Leaded 4140	187	72
Rycut-40	187	64
Nitralloy-135	243	47
Maxel-3	217	54
Hyten-3X	223	55
Hyten-B	187	60
81B40	179	60
Stressproof	265	50
52100	190	46

BASIC NUMBERS FOR VARIOUS GRADES OF CARBON ALLOY STEELS

1000	Plain carbon steels
1100	Resulfurized carbon steels
1300	Manganese 1.75%
2300	Nickel 3.50%
2500	Nickel 5.00%
3100	Nickel 1.25%, chromium 0.65-0.80%
3300	Nickel 3.50%, chromium 1.55%
4000	Molybdenum 0.25%
4100	Chromium 0.95%, molybdenum 0.20%
4300	Nickel 1.80%, chromium 0.50-0.80%, molybdenum 0.25%
4600	Nickel 1.80%, molybdenum 0.25%
4800	Nickel 3.50%, molybdenum 0.25%
5000	Chromium 0.30-0.60%
5100	Chromium 0.70-0.90%
6100	Chromium 0.90%, vanadium 0.10%
8100	
8600	NE (chrome-nickel-molybdenum)
8700	low alloy steels
9400	
14B00	
50B00	Boron treated steels
81B00	

Fig. 2-9.—Examples of machinability data and ratings for various steels related to AISI B1112 CD. (Registered trademarks: Rycut and Nitralloy—Joseph T. Ryerson & Son, Inc.; Maxel—Crucible Inc. division of Colt Industries; Hyten—Wheelock, Lovejoy & Co., Inc.; Stressproof—La Salle Steel Co.) *(Courtesy, Monarch Machine Tool Company, 10)*

1.1 TURNING

SINGLE POINT AND BOX TOOLS

MATERIAL	HARD-NESS BHN	CONDITION	DEPTH OF CUT in.	HIGH SPEED STEEL TOOL			CAST ALLOY TOOL		CARBIDE TOOL			
				SPEED fpm	FEED ipr	TOOL MATERIAL	SPEED fpm	FEED ipr	SPEED - fpm		FEED ipr	TOOL MATERIAL
									BRAZED	THROW-AWAY		
FREE MACHINING ALLOY STEELS (cont.)	200 to 250	Hot Rolled, Normalized, Annealed or Cold Drawn	.150	90	.015	M2, T5	100	.015	360	450	.015	C-6
			.025	130	.007	M2, T5	145	.007	440	550	.007	C-7
	275 to 325	Quenched and Tempered	.150	65	.015	M2, T5, T15	75	.015	300	375	.015	C-6
			.025	85	.007	M2, T5, T15	95	.007	385	475	.007	C-7
	325 to 375	Quenched and Tempered	.150	55	.015	M2, T5, T15	65	.015	225	275	.015	C-6
			.025	70	.007	M2, T5, T15	80	.007	290	365	.007	C-7
	375 to 425	Quenched and Tempered	.150	40	.010	T15, M41, M42, M43, M44	50	.010	175	215	.015	C-6
			.025	55	.005	T15, M41, M42, M43, M44	65	.005	225	280	.007	C-7
Resulphurized 3140 4150 4140 8640	45R$_c$ to 48R$_c$	Quenched and Tempered	.150	30	.010	T15, M41, M42, M43, M44	35	.010	160	200	.010	C-7
			.025	40	.005	T15, M41, M42, M43, M44	45	.005	185	225	.005	C-8
	48R$_c$ to 50R$_c$	Quenched and Tempered	.150	25	.010	T15, M41, M42, M43, M44	30	.010	120	150	.010	C-8
			.025	35	.005	T15, M41, M42, M43, M44	40	.005	140	175	.005	C-8
	50R$_c$ to 52R$_c$	Quenched and Tempered	.150	20	.010	T15, M41, M42, M43, M44	25	.010	100	125	.010	C-8
			.025	30	.005	T15, M41, M42, M43, M44	35	.005	125	150	.005	C-8
	52R$_c$ to 54R$_c$	Quenched and Tempered	.150	-	-	-	-	-	75	90	.010	C-8
			.025	-	-	-	-	-	85	105	.005	C-8
	54R$_c$ to 56R$_c$	Quenched and Tempered	.150	-	-	-	-	-	60	70	.010	C-8
			.025	-	-	-	-	-	75	90	.005	C-8
Leaded 41L30 51L32 41L40 86L20 41L47 86L40 41L50 52L100 43L47	150 to 200	Hot Rolled, Normalized, Annealed or Cold Drawn	.150	125	.015	M2, T5	145	.015	400	500	.015	C-6
			.025	170	.007	M2, T5	190	.007	475	600	.007	C-7
	200 to 250	Hot Rolled, Normalized, Annealed or Cold Drawn	.150	105	.015	M2, T5	120	.015	360	440	.015	C-6
			.025	145	.007	M2, T5	165	.007	435	550	.007	C-7

Fig. 2-10.—Example of machinability data taken from the *Machining Data Handbook*. The data are considered adaptable to both N/C and conventional machining methods (9).

Table II-12. IBM Division Standard—

Material	Machinability Classification	Speeds*							
		HSS Cutter				Carbide Cutter			
		Depth from .250 & over	Depth from .125 incl. .250	Depth from .062 incl. .125	Depth up to & incl. .062	Depth from .250 & over	Depth from .125 incl. .250	Depth from .062 incl. .125	Depth up to & incl. .062
Aluminum and Magnesium	Superior	1300	1350	1500	1500	2250	2500	3000	3000
	Excellent	1100	1150	1300	1300	1800	1850	2000	2000
	Good	1000	1050	1100	1100	1600	1650	1800	1800
	Fair	800	850	1000	1000	1400	1450	1600	1600
Bronze, Brass, and Copper	Free	425	500	600	750	930	1200	1500	1500
	Medium	240	270	350	420	450	600	775	900
	Hard	150	200	240	280	375	475	590	780
Malleable and Cast Iron	Excellent	150	180	225	275	410	520	630	800
	Good	120	150	180	225	320	390	490	590
	Fair	100	125	150	190	240	290	360	470
Carbon and Alloy Steels	Superior	230	280	340	410	590	700	850	1000
	Excellent	205	250	300	375	500	600	750	900
	Free	145	170	200	250	410	470	560	700
	Medium High	135	150	180	225	400	450	520	630
	Medium Low	80	100	130	160	240	280	350	410
	Hard	50	60	80	100	210	225	275	350
Sintered Metal	Free	350	420	500	610	1200	1600	2000	2000
	Medium High	150	180	225	280	580	750	900	1200
	Medium Low	125	150	190	240	380	490	600	790
	Hard	80	100	130	150	275	375	450	550
Stainless Steel	Free	100	115	135	175	250	260	310	400
	Medium	70	85	105	125	210	225	250	310
	Hard	45	60	70	90	160	180	210	260
Plastics Normal-lam. Asbes. and Gloss Thermosetting	Excellent	700	750	800	1000	1200	1400	1500	1500
	Hard	150	175	200	250	550	600	700	810
	Medium	360	425	500	600	1200	1300	1400	1400
Thermoplastic	Good	360	425	500	600	650§§	700§§	800§§	800§§
	Fair	250	300	400	490	250§§	300§§	400§§	490§§

*All speeds are in surface feet per minute (sfpm). Speeds apply to all types of cutters except end mills and Woodruff Keyseat Cutters. For super HSS tools (T-15, etc.) use 1.6 times the recommended sfpm for HSS.

†Indicates chip load per tooth for vertical and horizontal milling, duplex milling, and heavy milling under following conditions: over and including 100 A.A. surface finish and/or .002 in. and over tolerance on all vertical and horizontal mills and/or .003 in. tolerance on all duplex mills (when using both heads only).

‡A cutter is a helical mill only when its teeth have a helical angle of 25 deg to 52 deg and the cutter is over .750 in. wide.

§A form cutter is a cutter with form-relieved, undercut teeth which are ground on the tooth face.

Speeds and Feeds for Milling.

Face Mill	Shell Mill	Helical Mill‡ 25 deg, 45 deg & 52 deg	Angle Plain Slot and Side Mill	Stag. Tooth Slot Mill	Form Cutters§ Cor-ner‖	Form Cutters§ Regu-lar	Slitting Saw .010 to .030 wide#	Slitting Saw .032 to .060 wide**	Plain Tooth Saw††	Stag. Tooth Saw‡‡	Wood-ruff Keyseat Cutter
.015	.009	.013	.008	.0105	.009	.006	.0012	.003	.0045	.006	.003
.015	.009	.013	.008	.0105	.009	.006	.0012	.003	.0045	.006	.003
.012	.007	.011	.007	.0085	.007	.005	.001	.0024	.004	.005	.0025
.012	.007	.011	.007	.0085	.007	.005	.001	.0024	.004	.005	.0025
.015	.009	.013	.008	.0105	.009	.006	.0012	.003	.0055	.006	.003
.012	.007	.011	.007	.0085	.007	.005	.001	.0024	.0035	.005	.0025
.008	.005	.007	.0045	.0055	.005	.003	.0006	.0015	.0025	.003	.0015
.010	.006	.009	.0055	.007	.006	.004	.0008	.002	.003	.004	.002
.014	.0085	.010 .013	.008	.010	.008	.0055	.001	.003	.004	.0055	.003
.008	.005	.007	.0045	.0055	.005	.003	.0006	.0015	.0025	.003	.0015
.011	.007	.010	.006	.008	.0065	.0045	.0009	.0022	.0035	.0045	.0022
.011	.007	.010	.006	.008	.0065	.0045	.0009	.0022	.0035	.0045	.0022
.010	.006	.009	.0055	.007	.006	.004	.0008	.002	.003	.004	.002
.009	.006	.008	.005	.006	.0055	.0035	.0007	.0018	.003	.0035	.002
.008	.005	.007	.005	.0055	.005	.003	.0006	.0015	.0025	.003	.0015
.007	.004	.006	.004	.005	.004	.003	.0006	.0014	.002	.0025	.001
.0057	.005	.0057	.005	.0052	.0052	.004	.0012	.003	.0055	.0057	.002
.0057	.005	.0057	.005	.0052	.0052	.004	.0012	.003	.0055	.0057	.002
.0047	.0045	.0047	.0045	.0045	.0045	.003	.0006	.0015	.0025	.003	.001
.0047	.0045	.0047	.0045	.0045	.0045	.003	.0006	.0015	.0025	.003	.001
.008	.005	.007	.0045	.006	.005	.003	.0006	.0015	.0025	.003	.0015
.007	.0045	.006	.004	.005	.004	.003	.0006	.0014	.002	.0025	.001
.008	.005	.007	.0045	.006	.005	.003	.0006	.0015	.0025	.003	.0015
.007	.0045	.006	.004	.005	.004	.003	.0006	.0014	.002	.0025	.0015
.013	.0075	.0012	.007	.009	.008	.005	.001	.0025	.004	.005	.0025
.010	.006	.009	.0055	.007	.006	.004	.0008	.002	.003	.004	.002
.009	.006	.008	.005	.006	.0055	.0035	.0007	.0018	.003	.0035	.002
.012	.007	.011	.007	.0085	.007	.005	.0011	.0024	.004	.005	.0025

‖A corner form cutter is a cutter which cuts a radius on a sharp corner.

#A slitting saw .010 in. to .030 in. wide is a saw of the defined width with no side clearance.

**A slitting saw .032 in. to .060 in. wide is a saw of the defined width with sides ground concave to recess.

††A plain tooth saw is a cutter which is from .062 in. to and including .375 in. wide with teeth made parallel to the axis of the arbor.

‡‡A staggered tooth saw is a cutter which is from .188 in. to and including .250 wide with alternately staggered teeth.

§§Carbides are not recommended on thermoplastics because of good HSS results and the high angles recommended.

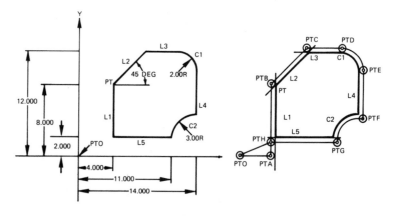

Fig. 2–11.—Sample part (left) and cutter path (right) drawings for a part to be programmed using ADAPT (12).

At this point of tape tryout, all corrections and all shop recommendations and standards may be incorporated into the master control tape. The master control tape is then converted into the commonly used, plastic-coated aluminum tape which reflects those program changes.

Manual Programming

Up to this point, most of the discussion has had a direct bearing on computer-aided programming methods and will be particularly inportant as N/C machining operations progress to direct computer control. Manual programming does not suffer the same problems. In manual programming, feeds, speeds, tool materials, and other machining variables are established at the discretion of the programmer for the particular machine operation. The programmer and the operator may change the tooling and the machine settings as the operation demands. As the level of sophistication of the machine, control system, and programming system increases, this flexibility is not desirable because such changes in the operation interfere with the total automatic control concept.

Adaptive Control

Adaptive control of N/C machine tools is being rapidly developed by the tool-building industry. In essence, the function of an adaptive control device is to sense and to revise machining conditions automatically in response to changes in other conditions. For example, variations in workpiece hardness are sensed by the machine, and spindle cutting speed is then modified to establish the predicted tool life. Such adaptive control devices as temperature, tool-wear, horsepower-requirement, and deflection sensors may be part of sophisticated adaptive control systems of the future. The signals generated by such devices may be entered into a computer which automatically responds with machining-variable changes of the correct direction and magnitude. Again, the best machinibility data available must be used to provide the adaptive control system with proper information upon which to act. Data systems for use with adaptive control are described in greater detail in Chapter 5.

PARTNO N/C 360 AD–APT SAMPLE PART PROGRAM	(1)
PTO = POINT/0, 0, 0	(2)
L1 = LINE/4, 0, 4, 8	(3)
PT = POINT/4.0, 8.0, 0	(4)
L2 = LINE/PT, ATANGL, 45	(5)
L3 = LINE/8, 12, 12, 12	(6)
L4 = LINE/14, 5, 14, 10	(7)
L5 = LINE/0, 2, 10, 2	(8)
C1 = CIRCLE/12, 10, 0, 2	(9)
C2 = CIRCLE/14, 2, 0, 3	(10)
INTOL/0	(11)
OUTTOL/.005	(12)
CUTTER/.25, .125	(13)
SPINDL/2000, CLW	(14)
COOLNT/ON	(15)
FEDRAT/20.0	(16)
FROM/PTO	(17)
GO/TO, L1	(18)
TLLFT, GOLFT/L1, PAST, L2	(19)
GORGT/L2, PAST, L3	(20)
GORGT/L3, TANTO, C1	(21)
GOFWD/C1, TANTO, L4	(22)
GOFWD/L4, PAST, C2	(23)
GORGT/C2, PAST, L5	(24)
GORGT/L5, PAST, L1	(25)
GOTO/PTO	(26)
COOLNT/OFF	(27)
SPINDL/OFF	(28)
FINI	(29)

Fig. 2–12.—ADAPT part program manuscript for the part shown in Fig. 2–11 (12).

60   TYPICAL DATA SYSTEMS FOR N/C PROGRAMMING

```
o                                                                        o
o                                                                        o
o    IBM SYSTEM/360 AUTOSPOT/AD-APT  INPUT TRANSLATOR. . .SECTION 1     05/15/70  PAGE 001   o
o    PARTNO/ ADAPT SAMPLE PART                                ASN   00001  o
     PTO = PCINT/0, 0, 0                                      ASN   00002
o    L1=LINE/4, 0, 4, 8                                       ASN   00003  o
     PT=POINT/4.0, 8.0,0                                      ASN   00004
o    L2 = LINE/POINT, ATANGL, 45                              ASN   00005  o
     L3 = LINE/8, 12, 12, 12                                  ASN   00006
o    L4 = LINE/14, 5, 14, 10                                  ASN   00007  o
     L5 = LINE/0, 2, 10, 2                                    ASN   00008
o    C1=CIRCLE/12, 10, 0, 2                                   ASN   00009  o
     C2=CIRCLE/14, 2, 0, 3                                    ASN   00010
o    INTOL/C                                                  ASN   00011  o
     OUTTOL/.005                                              ASN   00012
o    CUTTER/.25, .125                                         ASN   00013  o
     SPINDL/2000, CLW                                         ASN   00014
o    COOLNT/ON                                                ASN   00015  o
     FEDRAT/20.0                                              ASN   00016
o    FROM/PTO                                                 ASN   00017  o
     GO/TO, L1                                                ASN   00018
o    TLLFT, GOLFT/L1, PAST, L2                                ASN   00019  o
     GORGT/L2, PAST, L3                                       ASN   00020
o    GORGT/L3, TANTO, C1                                      ASN   00021  o
     GOFWD/C1 TANTO, L4                                       ASN   00022
o    GOFWD/L4, PAST, C2                                       ASN   00023  o
     GORGT/C2, PAST, L5                                       ASN   00024
o    GORGT/L5, PAST, L1                                       ASN   00025  o
     GOTO/PTO                                                 ASN   00026
o    COOLNT/OFF                                               ASN   00027  o
     SPINDL/OFF                                               ASN   00028
o    NOPOST, CLPRNT                                           ASN   00029  o
     FINI                                                     ASN   00030
o                                                                        o
```

Fig. 2-13.—Part program output listing for the part shown in Fig. 2–11. The processor lists the part program input data, and the CLPRINT command indicates that a CLFILE listing is requested (12).

DATA RECORD KEEPING

The data storage and access capabilities of a system play an important part in the development of a systems approach to processing and manufacturing, particularly in the N/C systems technique. In systems development, the master machine control tape may be used as an up-to-date machining information storage device. Incorporated in the control tape are shop recommendations and standards and accurate machinability information. The tape also represents the latest in engineering changes for the workpiece design. Manufacturing records of workpiece production, including part identification, machine-tool use, operator requirements, date of manufacture, previous production schedules, manufacturing time per piece, and other important production data, may be produced in printed form along with the master tape.

This manufacturing information, with the control tape, may be filed and stored for later use. Many N/C-oriented manufacturing installations find it advisable to keep accurate, up-to-date records of all N/C manufacturing experiences. Enterprising organizations have found it beneficial to abstract accurate and proven machinability information for the development of future machining standards and recommendations from the computer. Essentially, these standards and recommendations are derived from qualified N/C machining experiences, and they eventually become a library of machinability information available to the parts programmer. Thus, a full-cycle system is developed whereby the parts programmer enters approximate information which is in turn modified, proven, and found acceptable. This information, collected and produced in a suitable format, then returns to the parts programmer as accurate machinability data for future reference.

```
IBM SYSTEM/360 AUTOSPOT/AD-APT CLFILE PRINT AND TRACUT (SECTION 3)          03/15/70  PAGE 001
                                                                     ASN 00001       DISKFILE 00001
PARTNO/       /ADAPT SAMPLE PART                                                      DISKFILE 00002
                                                                     ASN 00011       DISKFILE 00003
INTOL/        .000000000 E+00                                                         DISKFILE 00004
                                                                     ASN 00012       DISKFILE 00005
OUTTOL/       .500000000 E-02                                                         DISKFILE 00006
                                                                     ASN 00013       DISKFILE 00007
CUTTER/       .250000000 E+00    .125000000 E+00                                      DISKFILE 00008
                                                                     ASN 00014       DISKFILE 00009
SPINDL/       .200000000 E+04 CLW                                                     DISKFILE 00010
                                                                     ASN 00016       DISKFILE 00011
COOLNT/       ON                                                                      DISKFILE 00012
                                                                     ASN 00016       DISKFILE 00013
FEDRAT/       .200000000 E+02                                                         DISKFILE 00014
                                                                     ASN 00017       DISKFILE 00015
FROM/POINT               NAME PTO                                                     DISKFILE 00016
        X              Y              Z
.000000000 E+00  .000000000 E+00  .000000000 E+00                    ASN 00018       DISKFILE 00017
GOTO/POINT               NAME L1                                                      DISKFILE 00018
        X              Y              Z
.387500000 E+01  .000000000 E+00  .000000000 E+00                    ASN 00019       DISKFILE 00019
GOTO/POINT               NAME L1                                                      DISKFILE 00020
        X              Y              Z
.387500000 E+01  .805177669 E+01  .000000000 E+00                    ASN 00020       DISKFILE 00021
GOTO/POINT               NAME L2                                                      DISKFILE 00022
        X              Y              Z
.794822330 E+01  .121250000 E+02  .000000000 E+00                    ASN 00021       DISKFILE 00023
GOTO/POINT               NAME L3                                                      DISKFILE 00024
        X              Y              Z
.118534580 E+02  .121250000 E+02  .000000000 E+00                    ASN 00022       DISKFILE 00025
DS IS CIRCLE             NAME C1                                                      DISKFILE 00026
        X              Y              Z              A              B              C              R
.120000000 E+02  .100000000 E+02  .000000000 E+00  .000000000 E+00  .000000000 E+00  .100000000 E+01  .200000000 E+01
                                                                     ASN 00022       DISKFILE 00027
GOTO/POINT               NAME C1                                                      DISKFILE 00028
        X              Y              Z
.119998180 E+02  .121299999 E+02  .000000000 E+00
.122908539 E+02  .121110483 E+02  .000000000 E+00
.125764342 E+02  .120505178 E+02  .000000000 E+00
.128512022 E+02  .119525252 E+02  .000000000 E+00
.131100039 E+02  .118179084 E+02  .000000000 E+00
.133479850 E+02  .116491926 E+02  .000000000 E+00
.135606815 E+02  .114495424 E+02  .000000000 E+00
```

- -

```
.137441038 E+02  .112227027 E+02  .000000000 E+00
.138948114 E+02  .109729283 E+02  .000000000 E+00
.140099775 E+02  .107049045 E+02  .000000000 E+00
.140874418 E+02  .104236586 E+02  .000000000 E+00
.141226538 E+02  .101585332 E+02  .000000000 E+00
.141250000 E+02  .999430423 E+01  .000000000 E+00
                                                                     ASN 00023       DISKFILE 00029
GOTO/POINT               NAME L4                                                      DISKFILE 00030
        X              Y              Z
.141250000 E+02  .487228132 E+01  .000000000 E+00                    ASN 00024       DISKFILE 00031
DS IS CIRCLE             NAME C2                                                      DISKFILE 00032
        X              Y              Z              A              B              C              R
.140000000 E+02  .200000000 E+01  .000000000 E+00  .000000000 E+00  .000000000 E+00  .100000000 E+01  .300000000 E+01
                                                                     ASN 00024       DISKFILE 00033
GOTO/POINT               NAME C2                                                      DISKFILE 00034
        X              Y              Z
.139554593 E+02  .487465494 E+01  .000000000 E+00
.136174290 E+02  .484943230 E+01  .000000000 E+00
.132847178 E+02  .478459986 E+01  .000000000 E+00
.129619497 E+02  .468105885 E+01  .000000000 E+00
.126536114 E+02  .454024859 E+01  .000000000 E+00
.123639893 E+02  .436612647 E+01  .000000000 E+00
.120971092 E+02  .415514076 E+01  .000000000 E+00
.118566811 E+02  .391619656 E+01  .000000000 E+00
.116460472 E+02  .365061541 E+01  .000000000 E+00
.114681364 E+02  .336208915 E+01  .000000000 E+00
.113254189 E+02  .305402856 E+01  .000000000 E+00
.112198817 E+02  .273250764 E+01  .000000000 E+00
.111529907 E+02  .240020417 E+01  .000000000 E+00
.111256758 E+02  .206233748 E+01  .000000000 E+00
.111326670 E+02  .187500000 E+01  .000000000 E+00
                                                                     ASN 00025       DISKFILE 00035
GOTO/POINT               NAME L5                                                      DISKFILE 00036
        X              Y              Z
.387500000 E+01  .187500000 E+01  .000000000 E+00                    ASN 00026       DISKFILE 00037
GOTO POINT               NAME PTO                                                     DISKFILE 00038
        X              Y              Z
.000000000 E+00  .000000000 E+00  .000000000 E+00                    ASN 00027       DISKFILE 00039
COOLNT/OFF                                                                            DISKFILE 00040
                                                                     ASN 00028       DISKFILE 00041
SPINDL/OFF                                                                            DISKFILE 00042
                                                                     ASN 00030       DISKFILE 00043
                              FINI                                                    DISKFILE 00044
                    END OF SECTION 3
```

Fig. 2–14.—Program output listing of all calculations required to control the cutter path shown in Fig. 2–11 (12).

OPTIMIZING N/C DATA AND FORMATS

Several N/C program formats are available. Each format is suited to a programming system and a machine tool. The parts programmer usually finds it relatively easy to understand and use a particular format. However, some problems may exist when a parts programmer is responsible for multiple formats and program systems within the same installation. In addition, the multiplicity of programming systems, language variations, and other complexities of program development require that the parts programmer carefully study the requirements of each. Admittedly, a programming system is not difficult to implement and maintain. Self-training, format training, and experience in the use of programming systems represent but a small problem to the user. Ideally, however, one standardized program-system language and format, applicable to all types of N/C operations, would be desirable.

REFERENCES

1. "Numerical Control: Tomorrow's Technology Today," *Proceedings of the Fifth Annual Meeting and Technical Conference of the Numerical Control Society, Philadelphia, April 3-5, 1968,* Princeton, N.J., Numerical Control Society, 1968.
2. *Numerical Contouring and Positioning Controls,* General Electric Company, Waynesboro, Va.
3. *N/C Handbook,* Detroit, The Bendix Corporation, Bendix Industrial Controls Division, 1967.
4. American Society of Tool and Manufacturing Engineers, *Numerical Control in Manufacturing,* New York, McGraw-Hill Book Company, Inc., 1963.
5. *Programmer's Manual No. 3—Cintimatic 200 Series Control, Publication No. M-2492-2,* Cincinnati, Cincinnati Milling Machine Co.
6. "Process Master for NAS 913, Rev. 2, August 15, 1965—Test 4.3.3.8.5, 3-Axis Profiler," Detroit, The Bendix Corporation, 1965.
7. *Part Processing Manual No. TGP-61, Milwaukee-Matic Model II,* Milwaukee, Kearney and Trecker Corporation.
8. Jack Thompson, "How to Improve Machining Center Productivity Through Better Programming," *Machinery,* Vol. 75, No. 5 (January, 1969).
9. *Machining Data Handbook,* 2d ed., Cincinnati, Metcut Research Associates, Inc., 1971.
10. *Speeds and Feeds for Better Turning Results,* 4th ed., Sidney, Ohio, Monarch Machine Tool Company.
11. *IBM Application Program—System/360 AD-APT/AUTOSPOT Numerical Control Processor (360A-CN-09X): Version 2, Part Programmer Manual,* White Plains, N.Y., International Business Machines Corporation, Technical Publications Department, 1967.

CHAPTER 3

COMPUTERIZED MACHINABILITY DATA SYSTEMS

Elbert J. Weller, General Electric Company
Charles Reitz, General Electric Company
Jacques Montaudouin, International Business Machines Corporation
Bernd E. Hirsch, Fokker GmbH
Horst Zölzer, EXAPT-Verein
W. Hans Engelskirchen, Systems, Science and Software
Emory W. Zimmers, Abex Corporation

The computerized systems for handling N/C machinability data presented in this chapter range from systems that are in everyday use to those which, at present, are being developed or are only ideas. These sophisticated, theorized systems are described to show how different system developers are approaching the problem of computerizing machinability data. All of these systems are in transition; they are growing and changing with the rapidly developing N/C metal-removal field, and other systems not listed here are also being evolved to keep pace with growing demand for computer aid in manufacturing.

Some of these systems may survive competition and some may not; some will be useful while others will not. But with the increasing number and variety of such systems which are available today, the user has only to decide which one is best for his own production purposes.

The computerized systems described in this chapter have many points in common with one another. Some simply place increased emphasis on certain of those common characteristics. For this reason, these descriptions may seem repetitive, but no attempt has been made here to consolidate them. It is necessary to give a complete picture of each system's capabilities—again, to allow the user to choose the capabilities that meet his own needs.

The General Electric data system, the first to be described, demonstrates the use of empirical machinability parameters in a mathematical computer model to determine optimum machining conditions for minimum cost and maximum production. The mathematical model, and the equations and factors used in its development, form the basis for nearly all computerized data systems.

The next system, the FAST system, shows how a computer may be used to simplify the lengthy process of selecting proper cutting tools for machining operations and of determining the proper feeds and speeds for those tools. FAST may also provide estimates of operation times.

The third system discussed, EXAPT, is an example of how the computer can be set up to find not only the optimum cutting tools for a particular operation, but also the correct operation sequences and cutting-tool movements.

Finally, the Abex system demonstrates the logic processes involved in the computer's calculations to find workable cutting conditions for minimum cost and maximum production.

In the last section of the chapter, the steps involved in the establishment and implementation of a computerized machinability data system are discussed.

The information on the systems presented here should be helpful to anyone interested in improved systems for computerizing machinability data. These systems can be used as starting points from which to develop larger, improved systems. The information on these present systems may also generate ideas for entirely new and better systems. In either case, the object here is to help people who are interested in coming up with new systems to reach the goal of an ultimate machinability data system or systems which will fully and accurately optimize machining conditions and costs.

GENERAL ELECTRIC DATA SYSTEM*

When machinability is analyzed from the systems viewpoint, electrical power, human effort, raw material, machine tools, and perishable cutting tools may be considered as inputs into the system, with machined products as the output. The main operating purpose of a manufacturing business is to control the inputs in order to produce a workpiece at either the minimum cost or the maximum production rate. There are definite operating conditions under which these two manufacturing objectives may occur. By the use of a digital computer, the parameters of an N/C machining system can be used to predict the operating conditions that are consistent with these objectives of manufacturing. Many parameters were considered as the General Electric data system was developed, and those of special interest will be elaborated on later in this section. These parameters are divided into the following four groups:

1) Workpiece parameters
   a) Material
   b) Microstructure
   c) Hardness
   d) Initial surface condition
   e) Surface finish required
   f) Starting geometric size and shape
   g) Final geometric size and shape
   h) Tolerances
2) Cutting-tool parameters
   a) Material
   b) Physical properties such as hardness, strength, wear resistance, crater resistance, microstructure, and chemistry
   c) Geometry, including side rake angle, back rake angle, top rake angle, nose radius, lead angle, and clearance angles

---

*The General Electric computerized machinability data system is a proprietary system of the General Electric Company. Requests for additional information concerning the system and its use should be addressed to the General Electric Company, Metallurgical Products Department, Detroit, Michigan.

    d) Coolant, either dry, oil, or emulsion
3) Machine-tool parameters
    a) Horsepower
    b) Spindle-speed limits
    c) Travel
    d) Rigidity
4) Operating condition parameters
    a) Tool-life speed
    b) Maximum-production speed
    c) Minimum-cost speed
    d) Feed
    e) Depth of cut.

Mathematical Model

The first step in using a computer to solve machinability problems is to prepare a mathematical model that includes all the significant parameters. When entered into the computer, the model will then serve as the basis for calculating the desired solution. The following formula is a mathematical model developed by Dr. W. W. Gilbert of the General Electric Company to compute cutting speed.

$$Speed = \frac{(CONST)\,(COOLF)\,(SURFF)\,(TMATF)\,(PROFF)\,(FLANK)^{.25}}{(K)^{1.72}\,(TLIFE)^{n}\,(FEED)^{.58}\,(DC)^{.2}}$$

$$\times \left(\frac{BHN_r}{BHN_w}\right)^{1.72} \times MR \qquad\qquad (3\text{--}1)$$

Where:

$CONST$	= A constant dependent upon basic tool material
$COOLF$	= Coolant factor based on coolant being used
$SURFF$	= Surface factor describing whether surface is clean, sand cast, or heat treated
$TMATF$	= Tool material factor
$PROFF$	= Profile factor—a function of nose radius, depth of cut, and cutting edge angle
$FLANK$	= Flank wear factor
K	= 160, the Brinell hardness number for the base material, AISI B1112 steel, used in establishing machinability ratings for steels
$TLIFE$	= Tool life
n	= Slope of the tool-life line
$FEED$	= Feed
DC	= Depth of cut
BHN_r	= Brinell hardness number at which the machinability rating was established
BHN_w	= Brinell hardness number of the workpiece
MR	= Machinability rating of workpiece material established at the hardness indicated

Data which substantiate this mathematical computer model, and determine the exponents used, are presented in the following subsections. These data are empirical data based upon the analysis of a broad background of shop and laboratory tests.

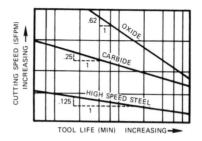

Fig. 3–1.—Tool life vs. cutting speed showing typical tool-life lines (2).

Tool Life. The most familiar relationship between cutting speed and tool life is the expression developed by F. W. Taylor (1):

$$VT^n = C \qquad (3\text{–}2)$$

Where:

V = Cutting speed in surface feet per minute (sfpm)
T = Time in minutes
n = Tool-life exponent
C = Constant representing cutting speed for a one-minute tool life

Taylor's expression can also be rewritten as:

$$\log V = C_1 - n \log T \qquad (3\text{–}2a)$$

When written in this logarithmic form, the expression can be plotted as a straight line on bilogarithmic graph paper. Fig. 3–1 shows a log-log plot of established test data for

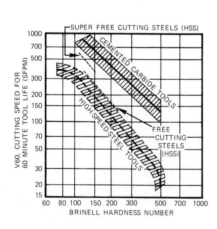

Fig. 3–2.—Log-log plot correlating cutting speed and Brinell hardness numbers for steels. *(Courtesy, Machining Development Operations, General Electric Company)*

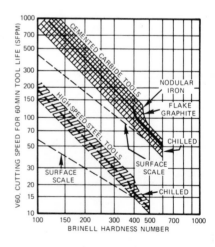

Fig. 3–3.—Log-log plot correlating cutting speed and Brinell hardness numbers for cast iron. *(Courtesy, Machining Development Operations, General Electric Company)*

which test results appear as straight lines. Laboratory tests reveal that the slopes of these lines, which provide the exponent n in Taylor's equation, are approximately .125 for high speed steel, .250 for cemented carbide, and .620 for cemented oxide. The tool-life parameter in the mathematical model is raised to the power of the same exponent, n.

Brinell Hardness. The bilogarithmic plot shown in Fig. 3–2 correlates cutting speed and the Brinell hardness for steel, indicating that this relationship is an inverse exponential function as is the relationship between cutting speed and Brinell hardness for cast iron shown in Fig. 3–3. The plot of machinability ratings against hardness numbers, as shown in Fig. 3–4, reveals an exponential relationship (3). Most published data indicates that cutting velocity varies inversely as an exponent of the Brinell hardness number. Laboratory data point to 1.72 as that exponent.

Fig. 3–5 shows two typical speed curves—one for B1112 steel having a machinability rating of 1 at 160 BHN, and the other for a material having a machinability rating

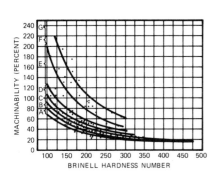

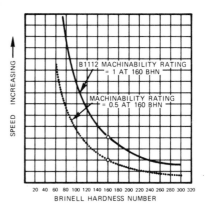

Fig. 3–4.—Machinability ratings vs. Brinell hardness numbers (3).

Fig. 3–5.—Speed curves for B1112 steel and another material. The machinability rating of a given material is the ratio of the speed indicated by its curve at a specified BHN to the speed of standard B1112 steel at 160 BHN (4).

of .5 at 160 BHN. The machinability rating of the second material is defined as the ratio of the speed shown on its curve at a specified BHN to the speed for B1112 at 160 BHN. Note that the second material has different machinability ratings at different Brinell hardness ratings. This variation in machinability is true for all materials. Microstructure is also extremely important to machinability, but it is usually difficult to obtain. BHN is a practical and usually reliable indicator of the variations in machining characteristics of a material.

Coolant or Cutting Fluid. Fig. 3–6 shows the results of studies into the effect of cutting fluid on the cutting-speed/tool-life relationship. Under certain conditions, extreme variation is possible. With dry cutting as the unit value, factors of 1.1 for oil and 1.25 for emulsion are used for the computer program. Computed speed will vary as these values are changed with respect to the cutting fluid used.

Surface Factor. Cutting speed and tool life are both affected by the original surface condition of the workpiece. Fig. 3–7 shows plots of cutting speed vs. tool life as related

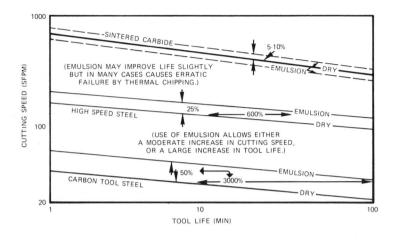

Fig. 3–6.—Results of studies of the effects of cutting fluids on the cutting-speed/tool-life relationship. *(Courtesy, Machining Development Operations, General Electric Company)*

to workpiece surface finish. Test data indicate a factor of 1.0 for a clean surface, .8 for a heat-treated surface, and .7 for a sand-cast surface.

Tool Material. Table III–1 indicates the results of years of experience in testing and use of different grades of carbide tool materials with a wide range of workpiece materials under various machining conditions. The selection guide represents a simplified tabulation of data used to determine the tool-material factors used in the mathematical model (6).

Table III–2 shows a partial list of the data stored in the material files. The first column is the identification number to be entered on the input data sheet. This number

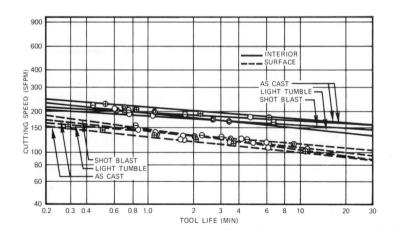

Fig. 3–7.—Cutting speed vs. tool life as related to workpiece surface finish (5).

is also used to determine the address on the disk file where this data is stored. The second set of data is the material name. This name is printed on the output sheet. The third column is the Brinell hardness at which the machinability rating shown in the fourth column was established.

Flank Wear. Fig. 3–8 shows a typical plot of flank wear as a function of time. This plot indicates that at a constant machining speed the flank wear land develops at

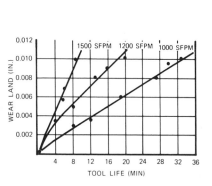

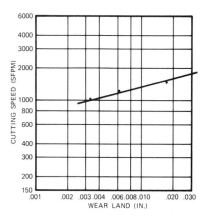

Fig. 3–8.—Flank wear as a function of time for various cutting speeds (4).

Fig. 3–9.—Log-log plot of speed vs. flank wear for constant time (4).

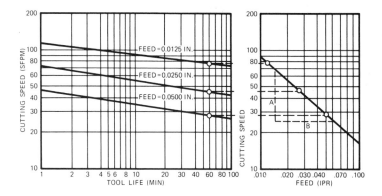

Fig. 3–10.—Effect of feed on cutting speed (7).

a constant rate and as a linear function of time. Fig. 3–9 illustrates a log-log plot of speed vs. the wear land for constant time. This curve is representative of data that has been used to determine .25 as the nominal slope of the carbide flank-wear curve used in the mathematical model.

Feed. The relationship of cutting speed to feed is shown in Fig. 3–10. The plot at the right (an interpolation of the family of feed curves) indicates that feed is an inverse

Table III-1. Starting Cutting Speeds and Insert Grade

Material			.002–.005 Feed .020–.050 Cut		.005–.010 Feed .050–.100 Cut	
			Speed (sfpm)	Grade	Speed (sfpm)	Grade
Steel—Average Work†						
Carbon	C1010–C1025		750	210	550	350,320
	C1030–C1095		600	210	440	350,320
Free Cutting	B1111–B1113		825	210	600	350,320
	C1106–C1151		800	210	590	350,320
Manganese	1320–1340		475	210	350	350,320
Nickel Chrome	3115–3150		475	210	350	350,320
Molybdenum	40XX–48XX		465	210	340	350,320
Chromium	5045–5152		475	210	350	350,320
	E50100–E52100		425	210	310	350,320
Cr Vanadium	6120–6152		450	210	330	350,320
Ni-Cr-Mo	86SXX–98XX		475	210	330	350,320
Ferrous Castings—Average Work‡						
Cast Iron	Hard	220/350 BHN	300	999	275	895,999
(No Alloy)	Medium	150/220 BHN	325	999	300	895,999
	Soft	100/150 BHN	350	999	325	895,999
Cast Iron	Hard	200/350 BHN	950	030	750	030
(No Alloy)	Medium	150/220 BHN	1200	030	900	030
	Soft	100/150 BHN	1800	030	1350	030
Cast Iron	Hard	270/600 BHN	275	210	250	350,320
(Alloy)	Medium	220/270 BHN	300	210	275	350,320
	Soft	160/220 BHN	325	210	300	350,320
	Chilled	600/750 BHN	45	999	30	895,999
Semi-	Up to 25%		325	210	300	350,320
steel	Over 25%		275	210	250	350,320
Malleable	Ferritic	110/150 BHN	350	210	325	350,320
Iron	Pearlitic	160/290 BHN	250	210	225	350,320
Nodular	Ferritic	140/210 BHN	375	210	350	350,320
Iron	Pearlitic	180/290 BHN	260	210	240	350,320
Stainless Steel						
AISI Type 300			300	999	285	895,999
AISI Type 400 and 500			375	210	325	350,320
Non-Ferrous Metals and Non-Metals—Average Work§						
Brass and	Hard		450	999	350	895,999
Bronze	Soft		550	999	450	895,999
	Free Cutting		850	999	600	895,999
Aluminum	Low Silicon		1500	999	1000	895,999
	High Silicon		700	999	500	895,999
Zinc Alloy	Die Casting		600	999	450	895,999
Rubber	Hard		600	999	475	895,999
	Soft		800	999	650	895,999
Copper			600	999	450	895,999
Monel[1]			275	999	250	895,999
Fibre			800	999	600	895,999
Plastic‖			800	999	600	895,999

Recommendations for Approximately 30 Min Tool Life (6).*

.010–.020 Feed .100–.200 Cut		.020–.040 Feed .200–.400 Cut		.040–.060 Feed .400–.600 Cut		.060–.100 Feed .600–1.000 Cut	
Speed (sfpm)	Grade	Speed (sfpm)	Grade	Speed (sfpm)	Grade	Speed (sfpm)	Grade
400	78,350	300	78B,350	175	370,78B	145	370
320	78,350	240	78B,350	150	370,78B	125	370
440	78,350	330	78B,350				
430	78,350	320	78B,350				
250	78,350	190	78B,350	150	370,78B	125	370
250	78,350	190	78B,350	150	370,78B	125	370
245	78,350	185	78B,350	95	370,78B	75	370
250	78,350	190	78B,350	100	370,78B	80	370
220	78,350	170	78B,350	90	370,78B	70	370
240	78,350	180	78B,350	95	370,78B		
250	78,350	190	78B,350	100	370,78B		
250	883,895	225	44A,883	115	44A,883	100	44A
275	883,895	250	44A,833	130	44A,883	110	44A
300	883,895	275	44A,833	160	44A,883	135	44A
225	78,350	200	78B,350				
250	78,350	225	78B,350				
275	78,350	250	78B,350				
275	78,350	250	78B,350				
225	78,350	200	78B,350				
300	78,350	275	78B,350				
210	78,350	200	78B,350				
325	78,350	300	78B,350				
225	78,350	210	78B,350				
225	883,895	215	44A,883				
275	78,350	225	78B,350				
275	883,895	225	883				
325	883,895	275	883				
450	883,895	350	883				
700	883,895	600	883				
350	883,895	300	883				
350	883,895	300	883				
400	883,895	350	883				
550	883,895	450	883				
350	883,895	250	883				
235	883,895	225	883				
450	883,895	350	883				
450	883,895	350	883				

Table III-1. Starting Cutting Speeds and Insert Grade Recommendations for Approximately 30 Min Tool Life (6). *(Continued)**

Material	.005–.010 Feed .050–.100 Cut		.010–.015 Feed .100–.015 Cut		.015–.020 Feed .150–.200 Cut	
	Speed (sfpm)	Grade	Speed (sfpm)	Grade	Speed (sfpm)	Grade
High Temperature Alloys						
René 41[2]	100	895	80	883,895	40	44A,883
J1500 (M252)	125	895	80	883,895	40	44A,883
J1570	125	895	89	883,895	40	44A,883
J1650	125	895	80	883,895	40	44A,883
A-286	150	895	80	883,895	50	44A,883
S-816	80	895	55	883,895	30	44A,883
Waspaloy[3]	125	895	80	883,895	40	44A,883
Udimet[4]	100	895	90	883,895	50	44A,883
Inconel X[5]	100	895	80	883,895	40	44A,883
Inconel 718[5]	100	895	80	883,895	40	44A, 883
Nimonic Alloys[6]	100	895	80	883,895	40	44A,883
Titanium (Pure)	450	895	225	883,895	100	44A,883
Titanium (Alloy) #	200	895	150	883,895	125	44A,883
J1300 (M308)	125	320	80	350	40	370
Timken[7] 16-25-6	175	320	90	350	50	370

*All grades shown are Carboloy (Reg. TM, Metallurgical Products Department, General Electric Company) cemented carbides. When two grades are shown, the first is generally the tougher; the second, in most cases, is more wear resistant and may give improved tool life when operating conditions permit.

†For "scale cuts," reduce speed approximately 20 percent.

‡For "as cast" surface, reduce speed approximately 20 percent.

§For rough work, reduce speed approximately 20 percent.

‖Thermosetting soft grades; for other plastics consult a carbide tool representative.

6Al-4V annealed. For other titanium alloys, consult a carbide tool representative.

[1]TM Huntington Alloy Products Division, The International Nickel Company, Inc.

[2]TM Vacuum Melted Alloys, Metallurgical Products Department, General Electric Company

[3]TM Special Metals Corporation

[4]TM Austenal Laboratories, Inc.

[5]TM Huntington Alloy Products Division, The International Nickel Company, Inc.

[6]TM Mond Nickel Company, Ltd.

[7]TM Timken Roller Bearing Company

exponential function (7). The exponent is determined to be .58; it represents the tangent of the angle of the speed/feed curve to the horizontal.

Fig. 3–11 shows the relationship between nose radius, theoretical surface finish, and feed (11). The expression used in the computer to find feed is as follows:

$$FEED = \left(\frac{21.6 \times nose\ radius \times RMS}{finish\ factor} \right)^{.5} \times 10^{-3} \qquad (3\text{--}3)$$

Where:

nose radius is expressed in inches

RMS = Surface finish expressed in microinches, root mean square

finish factor is an empirical value for different materials

Depth of cut. Fig. 3–12 shows the relationship between depth of cut and the cutting speed (7). This relationship is also an inverse exponential function in which the exponent is .2.

Table III-2. Typical Machinability Ratings.

Identification No.	Material	BHN	Machinability Rating
1	12% Chrome Stainless Iron	165	0.70
2	AISI 80B40	195	0.35
3	AISI 81B45	179	0.60
4	AISI 86B45	212	0.35
5	AISI 98B40	185	0.40
6	AISI 1020 (Casting)	134	0.60
7	AISI 1040 (Casting)	190	0.45
8	AISI 1330	223	0.60
9	AISI 3140	197	0.55
10	AISI 3250	220	0.45
11	AISI 3312	191	0.50
12	AISI 3340	220	0.45
13	AISI 3450	197	0.45
14	AISI 4130 (Casting)	175	0.35
15	AISI 4130	183	0.65
16	AISI 4140	190	0.55
17	AISI 4140 (Leaded)	187	0.70
18	AISI 4145	200	0.55
19	AISI 4340 (100% Pearlitic)	221	0.45
20	AISI 4340 (Spheroidized)	206	0.65
21	AISI 4340 (Casting)	300	0.25
22	AISI 4620	170	0.65
23	AISI 4640	187	0.55
24	AISI 4815	183	0.55
25	AISI 5120	191	0.65
26	AISI 6130	183	0.55
27	AISI 6135	190	0.55
28	AISI 6180	207	0.40
29	AISI 8030 (Casting)	175	0.45
30	AISI 8430 (Casting)	180	0.40
31	AISI 8620	194	0.60
32	AISI 8630	190	0.60
33	AISI 8630 (Casting)	240	0.30
34	AISI 8720	190	0.60
35	AISI 9255	218	0.45
36	AISI 9260	221	0.45
37	AISI 9262H	255	0.25
38	AISI A-286	300	0.10
39	AISI A-3115	160	0.65
40	AISI A-3120	150	0.65

Profile. The profile factor is charted in Fig. 3–13 as a function of nose radius, side cutting-edge angle, and depth of cut (9). These values are used in the mathematical speed model.

Optimum Speed

As stated earlier, it is necessary to control all the factors of the machinability system in order to produce machined products at either minimum cost or maximum production rate. Fig. 3–14 shows the total cost per piece as the sum of the load and idle cost,

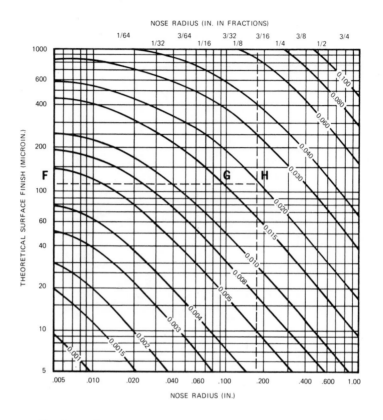

Fig. 3–11.—Relationship between nose radius, surface finish, and feed (8).

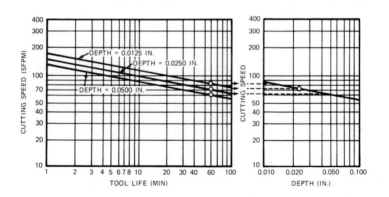

Fig. 3–12.—Relationship between depth of cut and cutting speed (7).

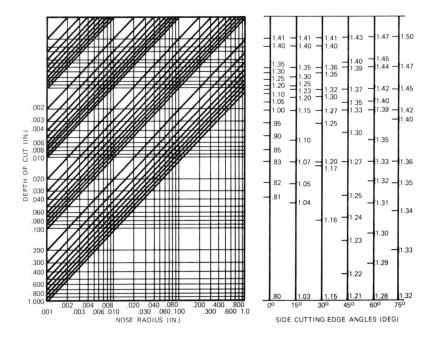

Fig. 3-13.—Profile factor as a function of nose radius, side cutting-edge angle, and depth of cut. Determine the point of intersection between NOSE RADIUS line and diagonal DEPTH OF CUT line. From intersection point, move horizontally to the right to the proper SIDE CUTTING-EDGE ANGLE line. Read tool profile factor from vertical scale (9).

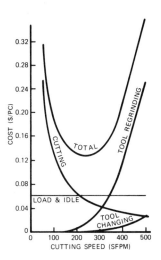

Fig. 3-14.—Costs per piece vs. cutting speed.

the cutting cost (tool in the cut), the tool changing cost, and the tool regrinding cost. Each of these costs is determined as follows (10):

1) Idle cost/piece = $K_1 \times$ idle time/piece $\qquad$ (3-4)

2) Cutting cost/piece = $K_1 \times$ cutting time/piece

$$= K_1 \frac{L\pi D}{12fV} \qquad (3\text{-}5)$$

3) Tool change cost/piece = $K_1 \times$ tool failures/piece $\times TCT$

$$= K_1 \frac{L\pi D(V)^{1/n-1}}{12f(C)^{1/n}} \times (TCT) \qquad (3\text{-}6)$$

4) Tool regrinding cost = $K_2 \times$ tool failures/piece

$$= K_2 \frac{L\pi D(V)^{1/n-1}}{12f(C)^{1/n}} \qquad (3\text{-}7)$$

Where:

K_1 = Direct labor rate plus overhead rate in dollars/min, including operator and helper labor, maintenance, power, depreciation, and insurance

K_2 = Tool cost per grinding, including original and regrinding costs in dollars/tool

L = Length of part in inches

D = Diameter of part in inches

V = Cutting speed in sfpm

f = Feed per revolution in inches

C = Cutting speed for one minute tool life

TCT = Tool-change time in minutes

Minimum Cost. After Eqs. 3-4, 3-5, 3-6, and 3-7 are added to obtain the total cost per piece, then the cutting speed for minimum cost may be found by differentiation in the following way (10):

Cutting speed for minimum cost/piece = $\dfrac{d(cost/piece)}{dV} = 0$

$$= 0 - K_1 \frac{L\pi D V^{-2}}{12f} + \left(\frac{1}{n} - 1\right) K_1 \frac{L\pi D(TCT)}{12f(C)^{1/n}} V^{1/n-2}$$

$$+ \left(\frac{1}{n} - 1\right) K_2 \frac{L\pi D}{12f(C)^{1/n}} V^{1/n-2}$$

$$K_1 V^{-2} = \left(\frac{1}{n} - 1\right) \frac{V^{1/n-2}}{C^{1/n}} (K_1 TCT + K_2)$$

$$V^{1/n} = \frac{C^{1/n}}{\left(\dfrac{1}{n} - 1\right)\left(\dfrac{K_1 TCT + K_2}{K_1}\right)}$$

$$V_{min.\,cost} = \frac{C}{\left[\left(\dfrac{1}{n} - 1\right)\left(\dfrac{K_1 TCT + K_2}{K_1}\right)\right]^n} \qquad (3\text{-}8)$$

Since $VT^n = C$ (Eq. 3-2), and tool life for minimum cost/piece =

$$T_{min.cost} = \left(\frac{1}{n} - 1\right)\left(\frac{K_1\ TCT + K_2}{K_1}\right) \tag{3-9}$$

Then cutting speed for minimum cost/piece may be converted to the following form:

$$V_{min.cost} = C\left(\frac{n}{1-n}\right)^n \left(\frac{K_1}{K_1\ TCT + K_2}\right)^n$$

$$= \frac{C}{T^n_{min.cost}} \tag{3-10}$$

Maximum Production Rate. Maximum production rate is calculated in the following way:

$$\text{Tool life for maximum production} = T_{max.prod.} = \left(\frac{1}{n} - 1\right) TCT \tag{3-11}$$

and

$$\text{Cutting speed for maximum production} = V_{max.prod.} = \frac{C}{\left[\left(\frac{1}{n} - 1\right) TCT\right]^n}$$

$$= \frac{C}{T^n_{max.prod.}} \tag{3-12}$$

High-Efficiency (Hi-E)* Machining

Hi-E is a technique of determining proper cutting speed for either maximum production or minimum cost. When cost/piece vs. speed is plotted on the same graph with pieces/time vs. speed, the range shown in Fig. 3-15 results. The Hi-E range is bounded on the left by the cutting speed at which minimum cost occurs (the lowest point on the cost curve) and on the right by the speed at which maximum production occurs (the highest point on the production curve; the point beyond which increasing tool-change time outweighs cutting-speed gains).

The optimum cutting speed will be found within the Hi-E range. The exact speed selected will depend on whether the emphasis is to be on cost or on production rate. Any speed within the Hi-E range will be a compromise between the two factors; any speed outside the Hi-E range will sacrifice both cost and production.

Because the Hi-E technique has been applied most commonly to turning operations, the following examples of the use of Hi-E are for axial turning. The technique has also been applied to milling operations, however, and it is reasonable to assume that it could be developed for other operations as well.

The basic Hi-E equation for tool life for maximum production is Eq. 3-11. The basic Hi-E equation for tool life for minimum part cost, Eq. 3-9, may be written as follows:

---

*The term *Hi-E* was originated by Thomas E. Hayes, service engineer of the Metallurgical Products Department, General Electric Company, in "How to Cut Costs with Carbides by 'Hi-E' Machining," *American Machinist*, Vol. 99, No. 27, pp. 121–23.

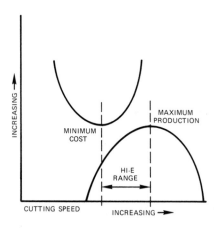

Fig. 3–15.—Hi-E range curves.

$$T_{min.\,cost} = \left(\frac{1}{n} - 1\right)\left(\frac{t}{M} + TCT\right) \qquad (3\text{–}9a)$$

Where:

$T_{min.\,cost}$ = Tool life (in the cut) for minimum part cost

n = Slope of the tool-life line

t = Total cost of the cutting edge, including tool-change cost, cutting-edge regrinding cost, and depreciation of brazed tools or mechanical holders

M = Machine labor and overhead rate

TCT = Tool-change time

Since $n = .250$ for carbide tools and $n = .125$ for HSS tools, then for carbide $(1/n - 1) = 3$ and for HSS $(1/n - 1) = 7$. The equations for carbide tools can be written as follows:

$$T_p = 3 \times TCT \qquad (3\text{–}13)$$

$$T_c = 3\frac{t}{M} + TCT \qquad (3\text{–}14)$$

The logic represented by these equations is included in the computer program.

Horsepower. The formula for determining horsepower requirements in axial-turning metal-removal applications is the following:

$$hp = \frac{\pi[(D_1)^2 - (D_2)^2] \times F \times RPM \times UHP \times C}{4E} \qquad (3\text{–}15)$$

Where:

D_1 = Original diameter

D_2 = Turned diameter

F = Feed in in./rev

RPM = Machine speed in revolutions per min

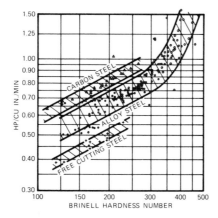

Fig. 3–16.—Relationship between unit horse-power *(UHP)* and Brinell hardness numbers (3).

E = Machine efficiency in percent

UHP = Horsepower requirements per cubic inch/min

C = Correction factor which has to be applied to compensate for the effect of feed on horsepower

Fig. 3–16 shows the relationship between *UHP* and Brinell hardness numbers. The two expressions used in the computer program are, for cast iron:

$$UHP = (BHN)^{1.9} \times ductility\,factor \times 33 \times 10^{-6} \qquad (3\text{--}16)$$

and for other materials:

$$UHP = (BHN)^{.5} \times ductility\,factor \times 615 \times 10^{-4} \qquad (3\text{--}17)$$

Using the Computer

With all the significant factors in the program accounted for, the computer can now be used to derive the following information for any given single-point turning problem:

1) Tool-life speed
2) Spindle speed
3) Horsepower required
4) Hi-E information
5) Recommended feed
6) Recommended grade of cemented carbide.

To find this information, the appropriate input data must be furnished to the computer in the correct form. Organization of the necessary information is simplified through the use of an input data sheet such as that shown in Fig. 3–17.

Most of the required input variables are self-explanatory, but the significance of some may be less obvious at first glance. The material number in the first section (Fig.

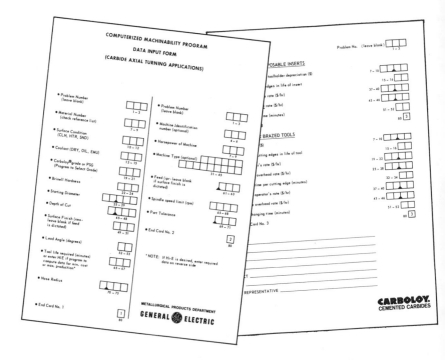

Fig. 3-17.—Input data sheet for use with the General Electric computerized machinability data system. (Carboloy is a registered trademark of the General Electric Company.) *(Courtesy, Metallurgical Products Department, General Electric Company)*

3-18), for example, is the identification number shown in Table III-2. The code number identifies the workpiece material and directs the computer to the record in the disk file where this material's characteristics are described. This information is used along with the Brinell hardness number to determine the material's machinability rating. Initial surface condition information is required because of its effect on tool life. Required surface finish is needed to compute feed; however, feed may also be dictated to the program (see Fig. 3-19).

Fig. 3-18 also shows how nose radius must be entered along with required surface finish whenever the computer is asked to compute feed. Lead angle is reflected in the profile factor, which affects feed rate. In this section, another system option is encountered—carbide tool grade selection.

If the program is to select the optimum grade of tool material to meet the problem requirements, the letters *PSG* (program select grade) are entered. If the carbide grade is to be dictated, the number designated for the grade is entered. An inherent danger in the latter option is that choice of a grade of tool material which is not suited to the application can produce an erroneous solution. To ensure the user's awareness of this danger, an appropriate footnote is printed along with the solution whenever the grade is dictated as shown in Fig. 3-20.

As shown in Fig. 3-19, available horsepower and spindle speed limit are significant in determining whether or not cutting can be done at the tool-life speed with the horse-

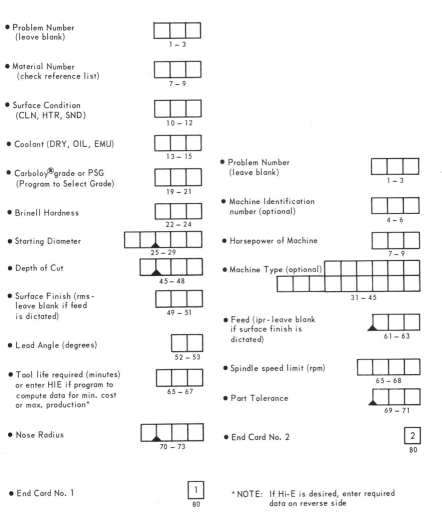

- Problem Number
  (leave blank)
  1 – 3

- Material Number
  (check reference list)
  7 – 9

- Surface Condition
  (CLN, HTR, SND)
  10 – 12

- Coolant (DRY, OIL, EMU)
  13 – 15

- Carboloy®grade or PSG
  (Program to Select Grade)
  19 – 21

- Brinell Hardness
  22 – 24

- Starting Diameter
  25 – 29

- Depth of Cut
  45 – 48

- Surface Finish (rms -
  leave blank if feed
  is dictated)
  49 – 51

- Lead Angle (degrees)
  52 – 53

- Tool life required (minutes)
  or enter HIE if program to
  compute data for min. cost
  or max. production*
  65 – 67

- Nose Radius
  70 – 73

- End Card No. 1
  1
  80

- Problem Number
  (leave blank)
  1 – 3

- Machine Identification
  number (optional)
  4 – 6

- Horsepower of Machine
  7 – 9

- Machine Type (optional)
  31 – 45

- Feed (ipr - leave blank
  if surface finish is
  dictated)
  61 – 63

- Spindle speed limit (rpm)
  65 – 68

- Part Tolerance
  69 – 71

- End Card No. 2
  2
  80

*NOTE: If Hi-E is desired, enter required data on reverse side

Fig. 3–18.—Detail of input data sheet— workpiece and cutting-tool data section. (Carboloy is a registered trademark of the General Electric Company.)

Fig. 3–19.—Detail of input data sheet— machine-tool data section.

power available on the machine tool. Final diameter tolerance determines what cutting-edge flank wear will be used to determine cutting speed in its turn. The closer the tolerance, the smaller will be the permissible flank wear. The correlation of diameter tolerance and flank wear is shown in Table III–3.

An outstanding feature of the present system is the option to call for a Hi-E computation. This is done by entering the letters *HIE* in the tool-life boxes on the first page of the input data sheet and by filling in all the boxes on page 2 under the section headed HI-E DATA FOR DISPOSABLE INSERTS (Fig. 3–21) or that headed HI-E DATA FOR BRAZED TOOLS (Fig. 3–22), depending upon the type of tools employed.

```
METALLURGICAL PRODUCTS DEPARTMENT
                    OF
              GENERAL ELECTRIC
      COMPUTERIZED MACHINABILITY RECOMMENDATION

WORKPIECE DATA

        MATERIAL                        AISI B-1112
        INITIAL SURFACE CONDITION       CLN
        BRINELL HARDNESS                160
        REQUIRED SURFACE CONDITION      250 MU IN. RMS
        STARTING DIAMETER               3.000 IN.
        FINAL DIAMETER                  2.500 IN.
        FINAL DIAMETER TOLERANCE        .005 IN.

MACHINE TOOL DATA

        NUMBER                          418
        OPERATION                       AXIAL TURN
        HORSEPOWER                      15
        SPINDLE SPEED LIMIT             1500 RPM

CUTTING TOOL DATA

        TOOL MATERIAL                   CEMENTED CARBIDE
        INPUT GRADE                     370
        NOSE RADIUS                     .045 IN.
        LEAD ANGLE                      6 DEG.

COOLANT                                 DRY

DYNAMIC DATA

        WEAR LAND                       .020 IN.
        RECOMMENDED FEED                .019 IPR
        TOOL LIFE INPUT                 15 MIN.
        TOOL LIFE SPEED                 662 FPM
        SPINDLE SPEED                   842 RPM
        H.P. REQUIRED                   34
        H.P. AVAILABLE                  15
        H.P. AVAILABLE SPEED            287 FPM
        H.P. AVAILABLE SPINDLE SPEED    365 RPM

NOTE: THIS SOLUTION IS BASED ON YOUR GRADE SELECTION.
```

Fig. 3–20.—Computer printout of information when cemented carbide tool grade is dictated. Note that this solution is based on the carbide grade selected.

Program Mechanics. Information from the input data sheet is key-punched into data-processing cards. The program, which is stored in the magnetic disk file, is directed into the main-frame memory with the program card. The data cards are then read by the program and the input information is checked by the computer to determine that the required parameters are present. If one or more needed pieces of information are missing, a warning message is typed out and the next set of data cards is read in.

If all required data are present, the workpiece material record is read from the disk-file storage into the main-frame memory. The machinability rating as related to the

Table III-3. Effects of Tolerance on Allowable Flank Wear.

Final Diameter Tolerance	Flank Wear
±.001	.005
±.002	.010
±.005	.020
greater than .005	.030

Problem No. (leave blank)

1 – 3

HI – E DATA FOR DISPOSABLE INSERTS

Price of insert plus toolholder depreciation ($)   7 – 10

Number of cutting edges in life of insert   15 – 16

Machine operator's rate ($/hr)   37 – 40

Machine overhead rate ($/hr)   43 – 46

Tool changing time (minutes)   51 – 52

End Card No. 3   80   3

Fig. 3–21.—Detail of input data sheet, page 2—Hi-E Data for Disposable Inserts section.

input hardness is determined and stored along with the work-material identification and grade of logic trail selector. The finish factor, the ductility factor, and an operation factor necessary for later use are determined.

Next, a series of calculations is made in which the following factors are determined:
1) Unit horsepower *(UHP)*
2) Surface condition factor
3) Depth of cut
4) Feed (if not specified)
5) Profile factor
6) Grade of tool material (if not specified)
7) Speed factor for the grade selected
8) Tool-life line slope as a function of tool grade
9) Coolant factor

The computer next tests to see if tool life was specified on the input data sheet. If so, factors necessary to evaluate speed are used in the routine, and the surface speed is

HI – E DATA FOR BRAZED TOOLS

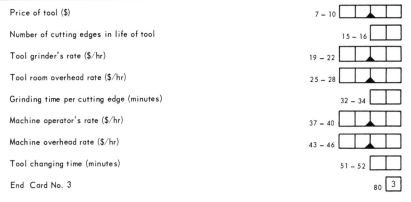

Price of tool ($)   7 – 10

Number of cutting edges in life of tool   15 – 16

Tool grinder's rate ($/hr)   19 – 22

Tool room overhead rate ($/hr)   25 – 28

Grinding time per cutting edge (minutes)   32 – 34

Machine operator's rate ($/hr)   37 – 40

Machine overhead rate ($/hr)   43 – 46

Tool changing time (minutes)   51 – 52

End Card No. 3   80   3

Fig. 3–22.—Detail of input data sheet, page 2—Hi-E Data for Brazed Tools section.

calculated. By using the calculated surface speed, the horsepower and the machine speed in revolutions per minute are determined. The tool-life horsepower is compared to the machine tool's available horsepower. If the equipment horsepower is greater than the tool-life horsepower, the solution is printed out. If, on the other hand, the equipment horsepower is less than that required to produce the exact tool-life speed, the machine speed possible with the available horsepower is computed.

When Hi-E is called for instead of a specified tool life, the computer calculates two sets of data. One set is for minimum cost and one is for maximum production. If available horsepower is less than either of the two calculated horsepower values, the maximum surface speed permissible with the equipment is computed. If available horsepower is adequate, the two sets of data for minimum cost and maximum production become part of the solution. When the problem is solved, the input information is printed out along with the solution. Fig. 3–23 shows a printout resulting from a tool-life input, and Fig. 3–24 shows a typical Hi-E printout.

METALLURGICAL PRODUCTS DEPARTMENT
OF
GENERAL ELECTRIC
COMPUTERIZED MACHINABILITY
RECOMMENDATION

WORKPIECE DATA

MATERIAL	AISI B-1112
INITIAL SURFACE CONDITION	CLN
REQUIRED SURFACE FINISH	250 MU. IN. RMS
BRINELL HARDNESS	160
STARTING DIAMETER	3.000 IN.
FINAL DIAMETER	2.500 IN.
FINAL DIAMETER TOLERANCE	± .005 IN.

MACHINE TOOL DATA

NUMBER	41A
OPERATION	AXIAL TURN
HORSEPOWER	15
SPINDLE SPEED LIMIT	1500 RPM

CUTTING TOOL DATA

TOOL MATERIAL	CEMENTED CARBIDE
RECOMMENDED GRADE	350
NOSE RADIUS	.045 IN.
LEAD ANGLE	6 DEG.

COOLANT   DRY

DYNAMIC DATA

WEAR LAND	.020 IN.
RECOMMENDED FEED	.019 IPR
TOOL LIFE INPUT	15 MIN.
TOOL LIFE SPEED	728 FPM
SPINDLE SPEED	926 RPM
H.P. REQUIRED	37
H.P. AVAILABLE	15
H.P. AVAILABLE SPEED	287 FPM
H.P. AVAILABLE SPINDLE SPEED	365 RPM

Fig. 3–23.—General Electric data system computer printout of information resulting from tool-life insertion.

METALURGICAL PRODUCTS DEPARTMENT
OF
GENERAL ELECTRIC
COMPUTERIZED MACHINABILITY
RECOMMENDATION

WORKPIECE DATA

MATERIAL	AISI C-1040
INITIAL SURFACE CONDITION	CLN
REQUIRED SURFACE FINISH	200 MU. IN. RMS
BRINELL HARDNESS	160
STARTING DIAMETER	3.000 IN.
FINAL DIAMETER	2.500 IN.
FINAL DIAMETER TOLERANCE	.005 IN.

MACHINE TOOL DATA

NUMBER	123
OPERATION	AXIAL TURN
HORSEPOWER	15
SPINDLE SPEED LIMIT	1500 RPM

CUTTING TOOL DATA

TOOL MATERIAL	CEMENTED CARBIDE
RECOMMENDED GRADE	370
NOSE RADIUS	250 IN.
LEAD ANGLE	5 DEG.

COOLANT	DRY

DYNAMIC DATA

WEAR LAND	.020 IN.
RECOMMENDED FEED	.036 IPR
MINIMUM MACHINING COST	
TOOL LIFE	39 MIN.
SPEED	312 FPM
SPINDLE SPEED	397 RPM
H.P. REQUIRED	27
MAXIMUM PRODUCTION RATE	
TOOL LIFE	14 MIN.
SPEED	404 FPM
SPINDLE SPEED	514 RPM
H.P. REQUIRED	35
H.P. AVAILABLE	15
H.P. AVAILABLE SPEED	168 FPM
H.P. AVAILABLE SPINDLE SPEED	213 RPM

Fig. 3-24.—General Electric data system computer printout of information resulting from a request for Hi-E data. Recommendations for minimum cost and maximum production rate are given.

FAST FOR FEED AND SPEED TECHNOLOGY

Because of the great number and variety of cutting tools available for N/C machining centers, and because of the characteristics of those machining centers as well, the calculations of optimum feed and speed for each type of cutting tool require not only experience and knowledge, but also a great deal of time. Today, computers are becoming more widely used to simplify the lengthy computations involved in supplying data for sophisticated N/C machines. Such programming systems as APT, ADAPT, and AUTOSPOT are typical computer programs developed to reduce the time and effort a parts programmer must expend.

Machinability data tables have also been compiled to supply data for computer use, but often such tables are voluminous sources that cover theoretical feeds and speeds without being specific about exact machining conditions. On the other hand, the computers themselves can be organized to analyze exact conditions, make tooling decisions, and communicate those decisions to the parts programmer in a fraction of the time that a highly trained and experienced engineer must devote to the job. The computer can list, update, and classify cutting tools in such a way that up-to-date tool records are available on request.

FAST (*Feed And Speed Technology*) is one programming system that will perform the cutting tool selection function. Written by N/C users for N/C users, FAST is written in the FORTRAN computer language and will run in the same computers that will handle the APT, ADAPT, and AUTOSPOT programming systems. The files from which data for use with the FAST program can be drawn are tailored to fit particular shop requirements based on the experiences of individual companies and their own methods of machine-tool operation. The program was written primarily to cover drilling, tapping, boring, center drilling, counterboring, countersinking, deep-hole drilling, coring, and milling operations performed by N/C machining centers.

Purpose

FAST was developed for three reasons:
1) To relieve the parts programmer of the lengthy, tedious process of selecting feeds and speeds for each tool used
2) To allow manufacturing time to be decreased by providing proper tools with the correct cutting-tool geometries
3) To ensure accuracy, flexibility, and the quick implementation of changes which result in the development of better tools and manufacturing methods.

Because of its data structure, the FAST program can be extended to perform two other functions in addition to determining feeds and speeds. It can also produce complete tool geometries for the cutting tools in question, and it can make accurate estimates of times for N/C operations. With the addition of these capabilities, three possible FAST data outputs, listed as follows and shown in Fig. 3–25, are available:
1) FAST—feeds and speeds of cutting tools only
2) FASTER (*Feed And Speed Test Estimating Random* access)—feeds, speeds, and geometries of cutting tools
3) FASTEST (*Feed And Speed Technology ESTimating*)—feeds, speeds, and estimated times of cutting operations.

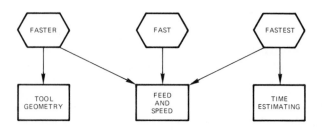

Fig. 3–25.—Data outputs of the FAST computer-aided machinability data systems.

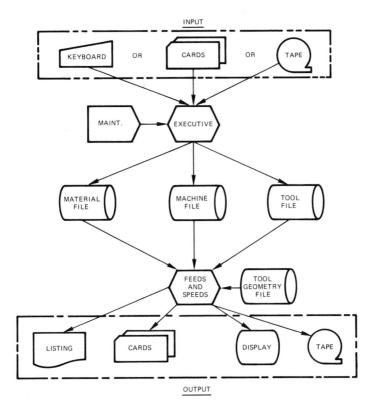

Fig. 3-26.—FAST general system flow chart.

Structure

FAST is made up of a main executive routine, a maintenance subroutine, and data files as shown in Fig. 3-26. Input data can be entered into the system through keyboard, punched cards, or magnetic tapes or disks. Output information is produced on printed listings, cards, magnetic tape, or a cathode ray tube (CRT) display.

The data files consist of four groups of information that influence the selection of feeds and speeds. These information groups are the following:

1) Material file
2) Machine file
3) Cutting-tool file
4) Tool geometry file.

Material File. The material file, or File 1, shown in Fig. 3-27, contains a complete description of the characteristics of the material used, including:

1) Material classification number
2) Machinability coefficient factor (the rating of relative machinability of the material)
3) Speeds and feeds for all types of tools, including three types of milling cutters.

MATERIAL LIST DATA

*MILL-1 FACE MILLS-FULL PACK-FLY CUTTERS-ETC
*MILL-2 SMALL END MILLS UP TO .500 INCH DIAMETER
*MILL-3 HELICAL MILLS-SHELL MILLS

MATERIAL CLASSIFICATION   7592
MACHINABILITY COEFFICIENT   0.55

	DRILL	BORE	REAM	SPDRL	CBORE	CSINK	MILL1*	MILL2*	MILL3*
HSS SFPM	60	80	45	60	70	70	150	80	185
FPT		0.0050			0.0040	0.0030	0.0100	0.0030	0.0080
CAR SFPM	120	300	180	120	290	290	400	340	350
FPT		0.0050			0.0040	0.0030	0.0120	0.0040	0.0080
FACTOR	1		-2	1					

MATERIAL CLASSIFICATION   7597
MACHINABILITY COEFFICIENT   0.50

	DRILL	BORE	REAM	SPDRL	CBORE	CSINK	MILL1*	MILL2*	MILL3*
HSS SFPM	40	65	40	40	60	60	100	75	130
FPT		0.0040			0.0030	0.0030	0.0070	0.0030	0.0080
CAR SFPM	80	300	120	80	225	225	350	300	250
FPT		0.0040			0.0030	0.0030	0.0080	0.0030	0.0060
FACTOR	1		-2	1					

MATERIAL CLASSIFICATION   8400
MACHINABILITY COEFFICIENT   0.35

	DRILL	BORE	REAM	SPDRL	CBORE	CSINK	MILL1*	MILL2*	MILL3*
HSS SFPM	25	40	15	25	30	30	55	40	50
FPT		0.0030			0.0010	0.0010	0.0060	0.0020	0.0060
CAR SFPM	50	160	45	50	130	130	200	200	200
FPT		0.0030			0.0010	0.0010	0.0100	0.0020	0.0080
FACTOR	1		-4	1					

MATERIAL CLASSIFICATION   14100
MACHINABILITY COEFFICIENT   3.00

	DRILL	BORE	REAM	SPDRL	CBORE	CSINK	MILL1*	MILL2*	MILL3*
HSS SFPM	225	300	150	225	225	225	450	300	450
FPT		0.0030			0.0050	0.0030	0.0050	0.0080	0.0080
CAR SFPM	500	500	300	500	475	475	1000	700	800
FPT		0.0030			0.0050	0.0030	0.0050	0.0080	0.0070
FACTOR	1		-4	1					

Fig. 3-27.—FAST material file listing.

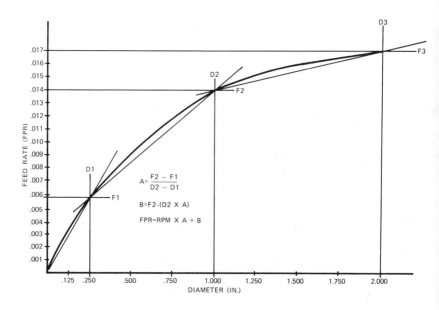

$$A = \frac{F2 - F1}{D2 - D1}$$

$$B = F2 - (D2 \times A)$$

$$FPR = RPM \times A + B$$

Fig. 3-28.—Interpolation of the feed vs. drill diameter curve to determine drilling feed.

MACHINE LIST DATA

MACH. GROUP	MAX RPM	MIN RPM	INC RPM	MAX. FEED	MIN. FEED	INC. FEED	MAX H.P.	MIN H.P.	COE
971	3125	150	79	38.5	0.5	0.1	5.0	2.5	1.00
1701	4000	100	10	50.0	2.0	1.0	10.0	2.5	0.80
1703	2070	100	32	50.0	1.0	1.0	5.0	5.0	1.00
1704	4000	100	10	50.0	1.0	1.0	10.0	2.5	1.00
1705	2400	60	32	98.0	0.1	1.0	10.0	10.0	1.00
2001	24000	8000	4	120.0	1.0	1.0	20.0	11.0	1.00
2003	5480	160	20	120.0	1.0	1.0	20.0	11.0	1.00
2006	5480	160	20	120.0	1.0	1.0	20.0	11.0	1.00

Fig. 3–29.—FAST machine reference file listing.

Feed per tooth is entered in the material data file in inches per tooth except for drilling, reaming, center drilling, and core drilling. In these four operations, a feed factor is used to determine feed per revolution based on tool diameter. Fig. 3–28 illustrates how feed is determined for a drill by interpolating the feed curve from three straight lines corresponding to the following diameter ranges:

1) To .250 in. diameter
2) .250 to 1.000 in. diameter
3) Over 1.000 in. diameter.

Machine File. The machine file, File 2, describes machine-tool parameters such as the following:

1) Maximum and minimum spindle speeds
2) Maximum and minimum feed rates
3) Maximum and minimum horsepower available at the spindle
4) A machine behavior coefficient factor.

The machine file is used only for quick reference during primary interrogation. It is completed by extensive, detailed machine-tool specifications. Fig. 3–29 illustrates the reference file, and the more detailed file is shown in Fig. 3–30.

—MACHINE TOOL SPECIFICATIONS

GROUP 1704

RPM RANGE	HP FORMULA
RP (1) = 590	HP (1) = RPM*0.0163 + 0.37
RP (2) = 1590	HP (2) = RPM*0.0065 - 0.33
RP (3) = 4000	HP (3) = RPM*0.0027 - 0.8

PROCESS TIME

POSITIONING FEED RATE	=150
ACCELERATION RATE	= 0
AUTO. TOOL CHANGE TIME	= 0.15
SHUTTLE TIME	= 0.2
POSITION, INDEX MIN. TIME	= 0.03
ADD TIME AFTER TL. CHANGE	= 0.16
ADD TIME AFTER TABLE INDEX	= 0.16
NUMBER OF TABLE	= 1
TABLE INDEX FACTOR	= 0.0003

Fig. 3–30.—Detailed FAST machine file listing.

NAME	TOOL CODE	TOOL DIAMETER	SETTING DISTANCE	S.S ROT	MILL SEL	TYPE	NO. OF TEETH	FLUTE LENGTH	DIA	EXTENSION LENGTH	HOLDER NUM	TOOL NUMBER	REMARK	EFFECTIVE LENGTH
CBORE	2704	0.4375	6.5000	0		HSS	3	0.0	0.0	0.0		763156146		
DRILL	2705	0.7500	7.5000	0		HSS	2	5.87	0.0	0.0		0		
DRILL	2706	0.5469	7.5000	0		HSS	2	0.0	0.0	0.0		0		
DRILL	2707	0.2380	6.5000	0		HSS	2	2.75	0.625	3.75		763016367		
DRILL	2708	0.1850	6.0000	0		HSS	2	2.31	0.625	3.50		763016297		
DRILL	2709	0.1719	6.0000	0		HSS	2	2.12	0.625	3.82		763016281		
DRILL	2710	0.0890	4.6000	0		HSS	2	1.26	0.625	3.12		763016165		
DRILL	2711	0.1015	5.0000	0		HSS	2	1.43	0.625	3.50		763016187		
DRILL	2712	0.2090	6.5000	0		HSS	2	2.50	0.625	4.00		763016337		
DRILL	2713	0.6250	7.5000	0		HSS	2	0.0	0.0	0.0		0		
DRILL	2714	0.0700	4.5000	0		HSS	2	1.00	0.625	3.25		763016129		
DRILL	2715	0.3750	7.0000	0		HSS	2	3.62	0.625	4.00		763016505		
DRILL	2716	0.2188	6.5000	0		HSS	2	2.50	0.625	4.00		763016345		
DRILL	2717	0.2900	7.0000	0		HSS	2	2.93	0.625	4.00		763016423		
REAM	2718	0.4385	7.5000	0		HSS	5	1.75	0.0	0.0		763522471		
DRILL	2719	0.4219	7.5000	0		HSS	2	3.93	1.000	3.50		763016535		
REAM	2720	0.1250	4.5000	0		HSS	3	0.87	0.625	2.50		763522126		
DRILL	2721	0.1160	5.0000	0		HSS	2	1.62	0.625	3.25		763016211		
REAM	2722	0.1260	4.5000	0		HSS	3	0.87	0.0	0.0		763522138		
REAM	2723	0.1875	5.0000	0		HSS	3	1.12	0.0	0.0		763522227		
REAM	2724	0.1865	5.0000	0		HSS	3	1.12	0.0	0.0		763522222		
REAM	2725	0.1885	5.0000	0		HSS	3	1.12	0.625	4.50		763522234		
REAM	2726	0.1562	4.5000	0		HSS	3	1.00	0.0	0.0		763522180		
DRILL	2727	0.1495	5.5000	0		HSS	2	1.87	0.625	3.50		763016249		
DRILL	2728	0.1800	6.0000	0		HSS	2	2.18	0.625	3.82		763016291		
MILL	2729	0.7500	7.5000	3		HSS	4	1.31	0.0	0.0		764019385		
REAM	2731	0.7500	7.5000	0		HSS	7	2.50	0.0	0.0		763522596		
REAM	2800	0.3760	7.0000	0		HSS	5	1.75	0.0	0.0		763522434		
DRILL	2801	0.3594	7.0000	0		HSS	2	3.50	0.625	3.50		763016491		
DRILL	2802	0.6875	7.5000	0		HSS	4	0.0	0.0	0.0			OCORE DRILL	
REAM	2803	0.2812	6.5000	0		HSS	5	1.50	0.0	0.0		763522353		
DRILL	2804	0.2770	6.5000	0		HSS	2	2.87	0.625	3.50		763016409		
DRILL	2805	1.0000	7.5000	0		HSS	4	0.0	0.0	0.0			OCORE DRILL	
DRILL	2806	0.2500	6.5000	0		HSS	2	2.75	0.625	3.50		763016381		
DRILL	2807	0.4130	7.0000	0		HSS	2	3.87	1.000	3.50		763016531		
CBORE	2808	0.5000	6.5000	0		HSS	3	0.0	0.0	0.0		763156178		
DRILL	2809	0.2969	6.5000	0		HSS	2	3.06	0.625	3.50		763016431		
DRILL	2810	1.0938	7.5000	0		HSS	2	0.0	0.0	0.0		0		
DRILL	2811	0.3125	6.5000	0		HSS	2	3.18	0.625	3.50		763016445		
DRILL	2812	1.5000	7.5000	0		HSS	2	0.0	0.0	0.0			OCORE DRILL	
DRILL	2813	1.5625	7.5000	0		HSS	2	0.0	0.0	0.0		0		
CBORE	2814	1.3125	6.0000	0		HSS	5	0.0	0.0	0.0		0		
DRILL	2815	1.2500	7.0000	0		HSS	2	0.0	0.0	0.0		0		

Fig. 3–31.—FAST tool file listing in numerical order by code number.

Tool File. File 3, the tool file, includes all data relative to cutting tools, cutting-tool holders, and adapters. Such information, as shown in Fig. 3–31, includes the following:

1) Cutting-tool name, identification code number, diameter, length, material composition, number of teeth, and flute length
2) Toolholder information
3) Adapter or extension diameter and length
4) Special information such as secondary diameter and length, tip angle, and remarks
5) Perishable tool identification code number.

Tool Geometry File. In addition to the three major FAST files for material, machine, and cutting tool, the FASTER system contains another file for tool geometry, illustrated in Fig. 3–32, which contains values for the computer equations used to generate proper tool-grinding geometries for the tools in question.

System Interrogation

The vocabulary for the FAST programming systems consists of major words, function names, and minor words. Major words control the major activities of the desired program. They are the following:

1) FAST
2) FASTER
3) FASTEST
4) LOAD
5) DELETE
6) ADD
7) PRINT

Function names control the data file from which information is to be taken. Function names are:

1) MATERIAL
2) MACHINE
3) TOOL
4) GEOMETRY

Finally, minor words control the machining operation for which data is to be selected. Minor words consist of the following:

1) TAP
2) REAM
3) BORE
4) DRILL
5) SPDRL (spot drill)

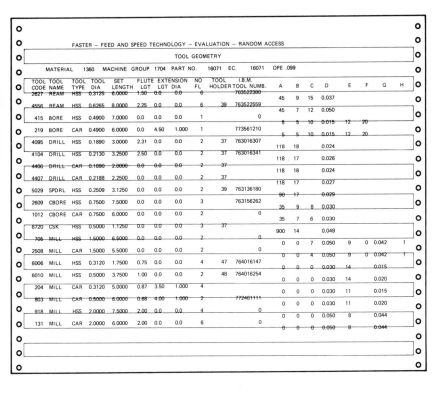

Fig. 3–32.—FASTER tool geometry file listing.

6) CBORE (counterbore)

7) CSK (countersink)

8) MILL

9) DEEPHOLE (deep-hole drill)

10) CORE (core drill)

The combination of any major word with a selected function name will allow working data to be recorded, modified, updated, or listed in or from the data-file names. For example, a combination of the word PRINT with the name TOOL in the form PRINTTOOL will result in a complete listing of the data in File 3 as shown in Fig. 3–31. This output listing shows tools classified in numerical order by their code numbers. Fig. 3–33 shows tools classified by diameter, type, and machining operation. File listings by material, machine, tool, or geometry are unique documents that can be used in many areas of parts programming, tool selection, and design.

TOOL LIST DATA

NAME	TOOL CODE	DIAMETER	SETTING DISTANCE	S.S ROT	MILL SEL	TYPE	NO. OF TEETH	FLUTE LENGTH	EXTENSION DIA	HOLDER LENGTH	NUM
MILL	2501	0.0400	4.5000	2		HSS	2	0.25	0.0	0.0	
MILL	419	0.0740	4.6250	2		HSS	2	0.25	0.0	0.0	
MILL	729	0.0937	6.5000	2		HSS	2	0.31	0.625	5.62	
MILL	614	0.1180	6.5000	2		HSS	2	0.37	0.625	5.25	
MILL	521	0.1200	4.5000	2		HSS	2	0.37	0.625	3.25	
MILL	2503	0.1200	6.5000	2		HSS	2	0.37	0.625	5.25	
MILL	2123	0.1250	4.5000	2		CAR	2	0.37	0.0	0.0	
MILL	1010	0.1250	5.5000	2		HSS	2	0.37	0.625	4.25	
MILL	6001	0.1400	1.5000	2		HSS	2	0.43	0.0	0.0	47
MILL	721	0.1440	6.5000	2		HSS	2	0.43	0.625	5.50	
MILL	731	0.1500	6.5000	2		HSS	2	0.43	0.625	5.50	
MILL	520	0.1560	4.5000	2		HSS	2	0.43	0.625	3.50	
MILL	312	0.1562	5.0000	2		HSS	2	0.43	0.625	4.00	
MILL	6002	0.1600	1.5000	2		HSS	2	0.43	0.0	0.0	47
MILL	2013	0.1735	4.5000	2		HSS	2	0.43	0.625	3.50	
MILL	529	0.1780	6.5000	2		HSS	2	0.43	0.625	5.50	
MILL	931	0.1840	6.0000	2		HSS	2	0.43	0.625	5.00	
MILL	206	0.1860	6.0000	2		HSS	2	0.43	0.625	5.00	
MILL	927	0.1860	6.5000	2		HSS	2	0.43	0.625	5.50	
MILL	128	0.1870	4.0000	2		HSS	4	0.43	0.625	3.00	
MILL	2014	0.1875	4.5000	2		HSS	4	0.43	0.625	3.50	
MILL	802	0.1880	6.5000	2		HSS	2	0.50	0.625	5.50	
MILL	323	0.1975	5.0000	2		HSS	2	0.50	0.625	5.50	
MILL	128	0.1975	5.5000	2		HSS	2	0.50	0.625	4.50	
MILL	2122	0.2180	6.5000	2		CAR	2	0.75	0.0	0.0	
MILL	2118	0.2190	4.5000	2		HSS	2	0.50	0.0	0.0	
MILL	2504	0.2460	6.5000	2		HSS	2	0.50	0.625	5.00	
MILL	519	0.2460	4.5000	2		HSS	2	0.50	0.625	3.00	
MILL	2027	0.2500	4.5000	2		HSS	2	0.50	0.625	3.00	
MILL	6003	0.2500	1.5000	2		HSS	2	0.50	0.0	0.0	47
MILL	2016	0.2500	6.5000	2		HSS	2	0.50	0.625	5.00	
MILL	2518	0.2760	6.0000	2		HSS	2	0.68	0.625	3.50	
MILL	6004	0.2800	1.5000	2		HSS	2	0.50	0.0	0.0	47
MILL	2007	0.2970	6.0000	2		HSS	2	0.56	0.625	4.50	
MILL	6005	0.3120	1.5000	2		HSS	2	0.56	0.0	0.0	47
MILL	204	0.3120	5.0000	2		CAR	4	0.87	1.000	3.50	
MILL	407	0.3120	7.0000	2		HSS	2	0.56	1.000	5.50	
MILL	6006	0.3120	1.7500	2		HSS	4	0.75	0.0	0.0	47
MILL	401	0.3125	4.5000	2		HSS	2	0.56	1.000	3.00	
MILL	1002	0.3240	4.0000	2		HSS	2	0.56	0.0	0.0	
MILL	824	0.3380	4.5000	2		HSS	2	0.81	0.0	0.0	
MILL	318	0.3440	4.5000	2		HSS	4	0.56	0.0	0.0	
MILL	723	0.3650	6.5000	2		HSS	2	0.75	1.000	5.00	
MILL	203	0.3680	5.0000	2		CAR	4	0.56	0.0	0.0	
MILL	405	0.3700	4.5000	2		HSS	2	0.75	0.0	0.0	
MILL	1026	0.3730	5.0000	2		HSS	2	0.75	0.0	0.0	
MILL	2121	0.3750	6.5000	2		CAR	2	0.87	0.0	0.0	
MILL	528	0.3750	4.5000	2		HSS	2	0.75	0.0	0.0	
MILL	106	0.3750	4.0000	2		HSS	4	0.75	0.0	0.0	
MILL	6008	0.3750	1.7500	2		HSS	4	0.75	0.0	0.0	47
MILL	6007	0.3750	1.5000	2		HSS	2	0.56	0.0	0.0	47
MILL	512	0.3750	6.0000	2		HSS	2	0.56	1.000	5.00	
MILL	129	0.3750	7.0000	2		HSS	2	0.56	1.000	5.00	
MILL	1000	0.3950	4.5000	2		HSS	2	0.81	0.0	0.0	
MILL	825	0.4110	4.5000	2		HSS	2	0.81	0.0	0.0	

Fig. 3–33.—FAST tool file listing by tool diameter, tool type, and operation.

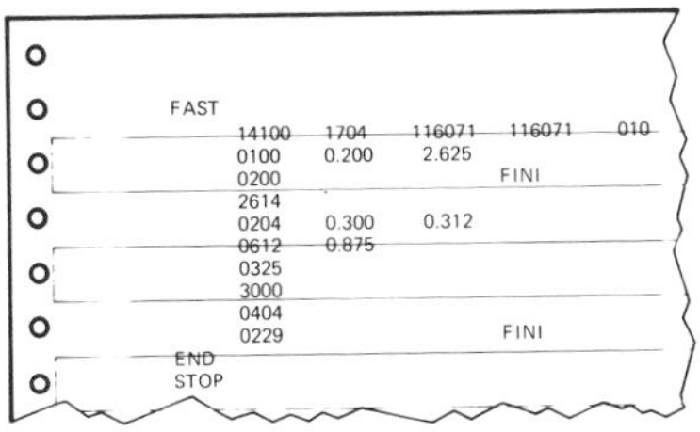

FAST				
14100	1704	116071	116071	010
0100	0.200	2.625		
0200			FINI	
2614				
0204	0.300	0.312		
0612	0.875			
0325				
3000				
0404				
0229			FINI	
END				
STOP				

Fig. 3–34.—Printout of typical input data for the FAST system.

FAST

The major word FAST triggers a complete technical analysis of the selected material, machine, or tool under the specific conditions required by the programmer. These conditions may be the depth and width of cut for the tools in question, the type of operation (whether roughing or finishing), and modifiers such as percentage of feed reduction.

Input. Fig. 3–34 shows a printout of a typical data input into the FAST system. The first line of the printout is a heading indicating material code number, machine group number, part number, engineering change (EC) number, and operation. The tool information is then listed on the lines following the heading.

Feeds and speeds of milling cutters are computed on the basis of required depth and width of cut, and when they are left blank by the parts programmer, FAST will assume and fill in the missing conditions. For example, if depth of cut is missing, the program will assume depth to be .062 in. (finishing cut). If width of cut is missing, the program will assume the width to be equal to tool diameter. If both depth and width of cut are missing, both the conditions given above will be assumed.

Other tools, such as drills, taps, and reamers, do not require such information. However, if conditions indicate, such tools will be analyzed to determine proper clearances, such as flute length vs. depth of cut, and feed and speed adjustments to compensate for depth/diameter ratios.

Output. Three output documents are produced by FAST analysis. The first document, shown in Fig. 3–35, is quite similar to the input data, but at this point some

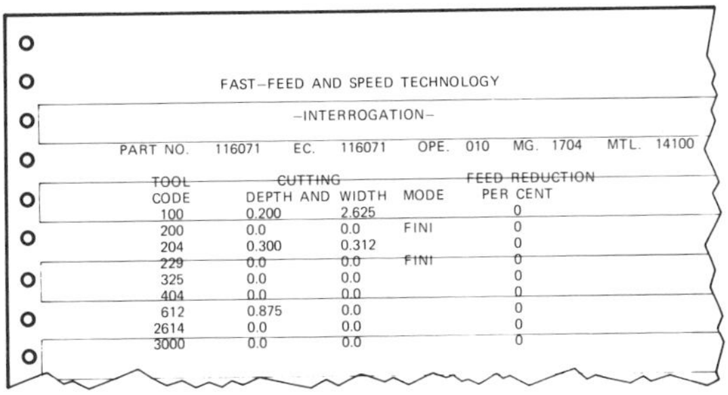

FAST–FEED AND SPEED TECHNOLOGY

–INTERROGATION–

PART NO. 116071 EC. 116071 OPE. 010 MG. 1704 MTL. 14100

TOOL CODE	CUTTING DEPTH	AND WIDTH	MODE	FEED REDUCTION PER CENT
100	0.200	2.625		0
200	0.0	0.0	FINI	0
204	0.300	0.312		0
229	0.0	0.0	FINI	0
325	0.0	0.0		0
404	0.0	0.0		0
612	0.875	0.0		0
2614	0.0	0.0		0
3000	0.0	0.0		0

Fig. 3–35.—FAST initial interrogation output document.

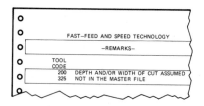

Fig. 3–36.—FAST initial interrogation remarks output document.

calculations have already been made. The material file has been interrogated for availability of the required material, and the material's parameters have been recorded in the common memory. The machine file, with its variables, has also been recorded in the memory. Finally, all tools have been checked for availability and have been resequenced by code number in ascending order, and the missing input parameters have been assumed.

Fig. 3–36 shows the remarks concerning the interrogation. This is the second output document.

The third output document is shown in Fig. 3–37 and is a complete listing of recommended feeds and speeds for each available tool. In this document the program will again produce messages indicating the various decisions made during computation. For example, the message "DEPTH OF CUT OF TOOL 204 ADJUSTED" warns the parts programmer that the original depth of cut of a particular tool was impractical or uneconomical.

Another possible set of input and output data is shown in Figs. 3–38 and 3–39, respectively. The output-data listing illustrates the feed-rate variations under different conditions. For tool 512, for example, the feed rate is 50 in./min. The feed rate for tool 6008, however, is 50 percent that for tool 512, or 25 in./min. The feed rate for tool 106 is reduced 25 percent to 37 in./min. Although no feed-rate reductions were requested for tools 2016, 2027, and 6003, the feed variations shown in Fig. 3–39 for

FAST–FEED AND SPEED TECHNOLOGY

–RECOMMENDED FEEDS AND SPEEDS–

MATERIAL 14100   MACHINE GROUP 1704 PART NO.   11607 E.C.   11607 OPE.   010

TOOL CODE	TOOL NAME	TOOL TYPE	TOOL DIA	SET LENGTH	FLUTE LGT	EXTENSION LGT DIA	NO FL	DEPTH OF CUT	WIDTH OF CUT	SPINDLE RPM FEED	TOOL HOLDER	I.B.M. TOOL NUMB.	REMARK
100	MILL	CAR	2.6250	6.0000	0.0	0.0 0.0	6	0.200	2.625	1447 43		0	FULL BACK
200	MILL	CAR	2.6250	4.0000	0.0	0.0 0.0	6	0.062	2.625	1664 29		0	FULL BACK
204	MILL	CAR	0.3120	5.0000	0.87	3.50 1.000	4	0.150	0.312	4000 35			
229	BORE	CAR	0.8750	6.0000	0.0	0.0 0.0	1	0.0	0.0	2497 4		0	DEVLIEG
404	BORE	CAR	0.8741	6.5000	0.0	0.0 0.0	1	0.0	0.0	2173 6		0	DEVLIEG
612	DRILL	HSS	0.1360	6.5000	1.75	4.25 0.625	2	0.875	0.0	4000 14		763016231	
2614	SPDRL	HSS	0.3750	7.5000	0.0	0.0 0.0	2	0.0	0.0	2279 2		0	
3000	TAP	HSS	250.2000	7.5000	1.00	0.0 0.0	2	0.0	0.0	199 9		766176223	1/4 20

NOTE. DEPTH OF CUT OF TOOL 204 ADJUSTED

Fig. 3–37.—FAST recommended feeds and speeds output document.

Fig. 3–38.—Tool interrogation input data printout for the FAST system.

those tools are the results of adjustments made after milling cutter strength was computed.

FASTER

The FASTER phase of the FAST program provides complete tool geometry in addition to tools, feeds, and speeds. The descriptions of the tools are based on the analyses of materials and tool diameters. Fig. 3–32 illustrates the data available through the FASTER program.

FASTEST

The FASTEST phase of FAST programming uses the steps of the FAST phase to calculate feed and speed for a particular tool. In addition to the machine-tool parameters listed in File 3, the FASTEST output contains complete specifications of machine constants such as the exact ranges of spindle speeds and feed rates, feed acceleration formulas, and tool-change index and shuttle-time variables. The results of these combined calculations provide not only accurate process times based on actual machining conditions, but also performed program checks. Printouts of FASTEST output data are shown in Figs. 3–40 and 3–41. Note that no matter what machine tool is investigated, FASTEST presents its data in unique language for that tool.

Fig. 3–39.—FAST recommended feeds and speeds for the tools interrogated in Fig. 3–38.

```
FASTEST
       14100   1704     116071    116071    010
       0100    0.200      2.625
       0200                        FINI
       2614
       0612    0.875
       3000
       0404
       0229                        FINI
END
1704
REMARK                                                 PART NO. 116071 E.C. 116071 OPE. 010
REMARK                                                 08 29 69 ESTIMATED BY J. MONTAUDOUIN
MILL       AUTO                13.00    0100           ROUGH MILL D.U.
MILL       AUTO     FINE        5.00    0200           FINI MILL D.U.
INDEX                         180.                     INDEX TO FACE 2
SPDRL      AUTO                0.125    2614      5    SPOT DRILL 250-20 HOLES
DRILL      AUTO                0.875    0612      5
TAP        AUTO                0.375    3000      3
TIME                           1.000                   CHECK 1.062 DIM. GAGE 1032541
INDEX                         90.                      INDEX TO FACE 3
BORE       AUTO                0.312    0404      2    ROUGH BORE (A) HOLES
BORE                           0.312    0229      2    FINI   "   "   "
INDEX                         90.                      INDEX TO FACE 0
SHUTTLE
END
STOP
```

Fig. 3–40.—FASTEST interrogation and output data printout.

EXAPT* SYSTEM

Development of the EXAPT program was begun in Europe about 1965 when a number of technical institutes and firms combined their resources to produce a data system that would automatically solve N/C machinability problems. The particular problems that were encountered were:

1) Selection of proper cutting tools
2) Calculation of tool paths with respect to workpiece geometry
3) Determination of the operation sequence
4) Determination of cutting conditions.

```
                    FASTEST–FEED AND SPEED TECHNOLOGY ESTIMATOR

MACHINE GROUP 1704  PARTS PER SET-UP          1

FUNCTION   TOOL    POSIT. DIST.   TOOL   NUMB.              REMARK                       TIME
           CHANGE                 CODE   HOLE
REMARK                                          PART NO. 116071 E.C. 116071 OPE. 010
REMARK                                          08 29 69 ESTIMATED BY J. MONTAUDOUIN
MILL       AUTO            13.00    100     0   ROUGH MILL D.U.                          0.459
MILL       AUTO    FINE     5.00    200     0   FINI MILL D.U.                           0.327
INDEX                     180.00      0     0   INDEX TO FACE 2                          0.286
SPDRL      AUTO             0.12   2614     5   SPOT DRILL 250-20 HOLES                  0.393
DRILL      AUTO             0.87    612     5                                            0.399
TAP        AUTO             0.37   3000     3                                            0.472
TIME                        1.00      0     0   CHECK 1.062 DIM. GAGE 1032541            1.000
INDEX                      90.00      0     0   INDEX TO FACE 3                          0.223
BORE       AUTO             0.31    404     2   ROUGH BORE (A) HOLES                     0.410
BORE                        0.31    229     2   FINI   "   "   "                         0.336
INDEX                      90.00      0     0   INDEX TO FACE 0                          0.223
SHUTTLE                     0.0       0     0                                            0.200

                          TOTAL TIME PER PALLET LOAD    4.73
                               TOTAL TIME PER PART      4.73
                               TOTAL TOOLS CHANGED      6
```

Fig. 3–41.—FASTEST output data including estimated operation times.

---

*The EXAPT computerized machinability data system is a proprietary system of EXAPT-Verein. Requests for additional information concerning the system and its use should be addressed to EXAPT-Verein, Postfach 587, Aachen, Germany.

EXAPT solves these problems by including both workpiece geometry and all technological operation data in its computations.

EXAPT is related to and interfaced with the APT programming system. Most of
the geometrical definitions of APT are included in EXAPT, but the latter contains a
more extensive technological capability. The EXAPT vocabulary is compatible with
APT language as far as possible, and further compatibility is constantly being developed by the interchange of ideas between APT and EXAPT managements. A parts
programmer can use either language, as needed, without difficulty; processor outputs
are identical no matter which program is used.

In an EXAPT part program, the geometrical and technological features of a part
are described with APT statements or with "APT-like" statements for features that are
defined in EXAPT but not in APT. The EXAPT processors then select operation
sequences, cutting tools, and collision-free tool motions. They also determine optimum
machining conditions from the economic, empirical, and theoretical metal-cutting conditions that are fed into them.

Language Parts

The complete EXAPT programming system consists of three language parts, listed
as follows:
1) EXAPT 1 deals with the programming of point-to-point and single straight-line
   path control. It is mainly used for drilling and simple milling.
2) EXAPT 2 is concerned with the programming of axial turning and contains
   both straight-path and contour control with circular and linear interpolation.
3) EXAPT 3 provides the capability for 2-1/2-dimensional contour milling.

Together, the three EXAPT language parts enable the parts programmer to describe
drilling, turning, and milling problems (11). The EXAPT 1 and 2 programming systems are operational and are being used in industry today. EXAPT 3 was introduced
in late 1970. Table III–4 shows the capabilities of the three EXAPT processors, and
Fig. 3–42 roughly shows the steps of industrial production with EXAPT from the
conception of a part to its manufacture (12).

In addition to the calculation of cutting conditions, EXAPT 1 automatically determines the premachining operations needed to accomplish the final operation as defined
in the part program. This capability is particularly important because of the great
variety of tools needed to completely manage a drilling job.

EXAPT 2 is being continually developed to automate and optimize the higher levels of processing in the planning stages. Although it still requires turning setup positions and a statement of the sequence of operations in the part program, area clearance
problems are automatically solved by continuous clearance checks. For this to be done,
however, the geometries of the workpiece blank and the finished part must be described
in the part program. Intermediate shapes and tool clearance movements are then
automatically calculated. EXAPT 3 also operates on this principle.

Input Data

As in the FAST programming system, the EXAPT system contain data files, to
which the computer has access, which store all the information on cutting tools,
machine tools, and materials. The part program must contain only that information
that varies from part to part; constant data from the files is usually supplied automatically by the computer. Fig. 3–43 shows a simplified flow chart for the EXAPT 2 sys-

Table III-4. Features of the EXAPT Programming System.

General Features	Arithmetic, Macros, Loops, Processor Control Statements		
	EXAPT 1	EXAPT 2	EXAPT 3
Geometric Definitions	point, line, circle patterns (arc, linear, random, mirror) transformations	point, line, circle contour description (blank and finished part)	point, line, circle patterns tabulated functions polygonal lines contours of pockets transformations
Technological Definitions	center drilling drilling counterboring reaming tapping milling (straight line) boring	for center holes: center drilling drilling counterboring reaming tapping straight-line turning contour turning grooving* thread cutting*	center drilling drilling counterboring reaming tapping boring straight-line milling contour milling area clearance milling for pockets
Executive Statements	callup of predefined machining operations callup of points and patterns simple execution statements	callup of predefined machining operations callup of coordinate of center hole callup of a part of the part contour which is to be machined simple execution statements	callup of predefined machining operations callup of point patterns callup of contour of pockets simple execution statements
Automatic Selection and Determination of:	tool path (Z coord.) cutting speed feed tools work sequence	for center holes: cutting speed feed tool path for turning operations: depth of cut cutting speed feed cutting distribution collision checking tools* work sequence*	depth of cut cutting speed feed tools cutting distribution path calculation collision checking
Fields of Application	hole-making operations simple straight-line milling	lathe work	2-1/2-axis milling hole-making operations

*Features under development

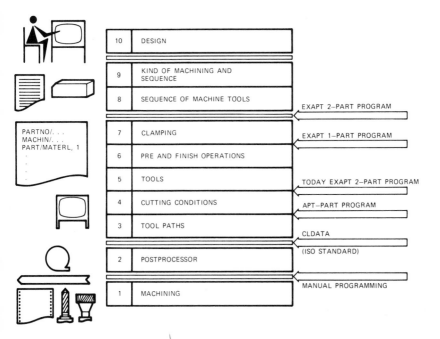

10	DESIGN		
9	KIND OF MACHINING AND SEQUENCE		
8	SEQUENCE OF MACHINE TOOLS		
		EXAPT 2–PART PROGRAM	
7	CLAMPING	EXAPT 1–PART PROGRAM	
6	PRE AND FINISH OPERATIONS		
5	TOOLS	TODAY EXAPT 2–PART PROGRAM	
4	CUTTING CONDITIONS	APT–PART PROGRAM	
3	TOOL PATHS	CLDATA	
2	POSTPROCESSOR	(ISO STANDARD)	
1	MACHINING	MANUAL PROGRAMMING	

PARTNO/. . .
MACHIN/. . .
PART/MATERL, 1

Fig. 3–42.—Steps in industrial computer-aided production with the EXAPT system.

tem. That for EXAPT 1 is nearly identical except that it contains routines for tool sequence and cutting tool selection in its technological processor. Note the processor's access to the data files in Fig. 3–43.

Data is entered into the files by a special form called an index card. An index card for materials is shown in Fig. 3–44 and one for turning tools in Fig. 3–45. Similar cards are used to enter information on other kinds of cutting tools and on machine tools. Thus, the EXAPT files contain all the information on the tools and materials that a particular plant uses in its operations—the "manufacturing capacity" of the organization. Updating routines allow easy maintenance of the files so that data can be constantly revised or added—when new tools or materials are used or when the company's economic goals change, for example.

Computer Processes

The mathematical equations used in the EXAPT technological processor describe the relationships of influential metal-cutting parameters. The equations that are selected and adapted should only meet workshop requirements, and numerical input values must be supplied. It is pointless to incorporate theoretical formulas in the program when they would have no noticeable effect on machine-tool setting values; this procedure would merely waste computer time.

Drilling. The cutting values for drilling are easier to determine than those for turning. Fig. 3–46 shows the two equations for feed rate and cutting speed for drilling. The cutting-speed value for each drilling operation can be taken from the card index

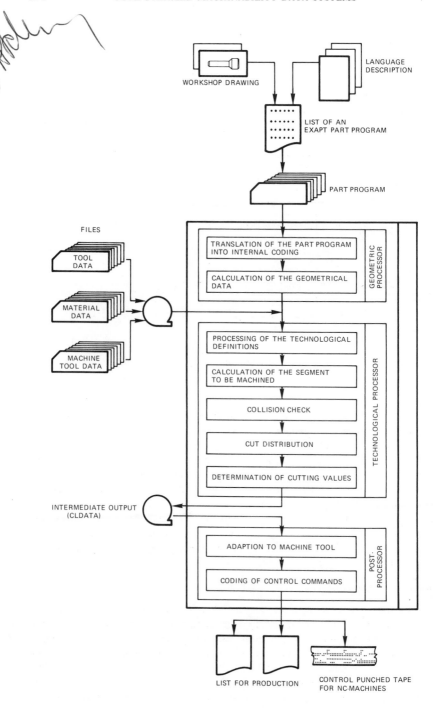

Fig. 3-43.—General EXAPT 2 processor flow chart.

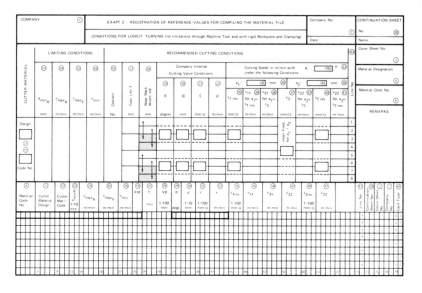

Fig. 3–44.—EXAPT materials file index card.

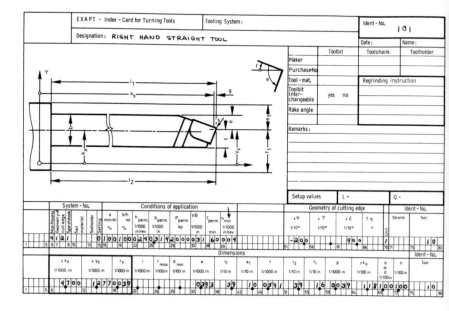

Fig. 3–45.—EXAPT cutting-tool file index card with tool data inserted.

data as a constant, but the feed rate must be more precise. The user must request feed rate based on drill diameter. The feed-rate/diameter relationship is shown in Figs. 3–28 and 3–46. The coefficients for the three straight-line segments are stored in the materials file.

The data contained in the materials file applies to tools with standard cutting-edge geometries. Drilling tools with cutting-edge geometries different from the standards can be used by including correction factors for feed and cutting speed on the applicable

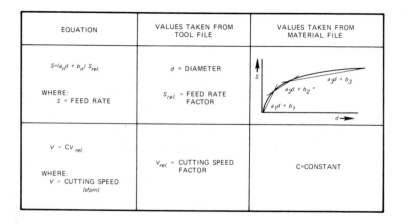

Fig. 3–46.—Calculation of feeds and speeds for EXAPT 1.

drilling-tool index card. The maximum permissible feed rate, which is also included on the index card, is the upper limit on the feed rate determined for the tool.

The technological processor can also generate commands for peck feeding and for changes of feed and speed in relation to hole depth. These practical methods for determining cutting values for drilling have met the needs of production satisfactorily.

Turning. Many more determining factors are used to obtain EXAPT 2 cutting values for turning operations (13, 14). For example, speed and feed must be constantly compared to limiting values for the machine tool, such as power, torque, and the available speed and feed range. In addition to tool material and tool radius for the tool shown in Fig. 3–45, other data are given in the "Conditions of Application" section of the tool index card. Fig. 3–47 shows the dimensions of tool geometry for turning tools which will be discussed below.

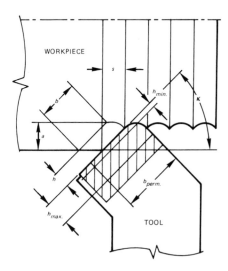

Fig. 3–47.—Turning-tool dimensions as entered in the EXAPT tool file—a, depth of cut; b, chip width; $b_{perm.}$, permissible engaged length of cutting edge; h, chip thickness; s, feed rate; $h_{min.}, h_{max.}$, permissible range of chip thickness; κ, approach angle.

The material file includes information for different part materials and, within each part material group, for different cutter materials (15). The processor automatically selects data for a particular part-material/cutter-material combination. Here again, the data is calculated for a standard tool, so the first two values under "Conditions of Application," $a_{min.rel.}$ and $b/h_{rel.}$, are included to correct the depth of cut and feed rate for tools that deviate from the standard. To correct for minimum depth of cut, $a_{min.}$ is taken from the material file based on the ductility of the part material, and the chip form relationship, b/h, is calculated with data from the material file for the cutting-edge geometry of the tool used. This geometry is usually different from that of the standard tool.

The values $h_{min.}$ and $h_{perm.}$ specify the permissible range of chip thicknesses and should be measured accurately, particularly if chip breakers are used. The usable length, $b_{perm.}$, indicates the permissible engaged length of the cutting edge of the tool.

The chip size of the chip cross-sectional area is controlled by the permissible cutting force, P_{perm}. When this value is estimated, shank cross-section, cantilevered length, condition of the cutting plate, and clamping should be considered. The value VB is a measurement of the wear permitted during the tool life, T_{perm}.

Tool Wear Control

Basically, tool wear is influenced by a number of factors which are still, in part, unexplained, and during chip removal tool wear can only be controlled by the selection of appropriate cutting conditions. For this reason, the relationships between wear and the cutting values must be known. Usually, the wear effect of cutting speed is greatest, the effect of feed is marked, and that of depth of cut is present but negligible. Taylor's expression, $VT^n = C$ (Eq. 3-2), adequately describes the tool-life/cutting-speed interdependence and is still widely used, although computers now provide the means of trying more accurate mathematical descriptions. In any case, feed-rate, depth-of-cut, and tool-wear parameters must be similarly considered as variables if their effect on cutting values is to be determined.

The width of the flank wear land is used as the measurement of wear because it can be assessed more easily than other wear characteristics, such as crater wear. However, crater wear behavior can be described by the same mathematical equations as flank wear, so both measurements may be used in the EXAPT system as long as the correct numerical values are provided in the equations.

All the tool-wear relationships can be combined in a single, multidimensional coordinate system for the following tool-life equation (16):

$$V = C \times a^E \times s^F \times T^G \times VB^H \qquad (3\text{-}18)$$

Where:

V	= Cutting speed (sfpm)
s	= Feed rate (ipr)
a	= Depth of cut (in.)
T	= Standard tool life (min)
VB	= Width of the flank wear land (in.)
C	= Constant
E, F, G, H	= Exponents

Weighting of the different factors in the equation is expressed by the numerical values of the exponents.

In order to obtain the highest possible yield with a tool that has been reground, the depth of cut chosen should be as large as possible. For great depths of cut, the number of cutting passes needed is reduced, thus saving not only cutting time but time for clearing, returning, and repositioning the tool. Feed rate and cutting speed have the same influence on cutting time. The smaller wear effect, and the advantage that specific cutting force decreases with increased chip thickness, favors an increased feed rate. However, these conditions occur only when additional limiting parameters are taken into account and when only cutting speed is varied to fulfil the tool-life equation.

Preset Tool-Life Values

Tool life is usually defined as the maximum service life of a cutting tool between two regrinding operations. Actually, the term implies that a certain quantity of cutting-tool material is consumed during tool life—in other words, that the tool has a certain "operational intensity."

The definition of tool life as the operational intensity of a tool is a valuable one for complex N/C machining processes. Experience with N/C machining centers shows that the empirical "handbook" values obtained for conventional machine tools must be reconsidered under the criteria of intensity of use. In order to completely machine a workpiece in one setup, for example, expensive cutting heads as well as cheap twist drills may be used on the same machining center for the same workpiece. The differences in wear behavior and price, however, make it necessary to determine tool-life values in the form of operational intensities, which may differ greatly from each other. Long tool life reduces down time for tool changing, but also requires the reduction of cutting values. On the other hand, short tool-life values permit high machining rates and reduced cutting times, but increase tool changing time. These factors must be balanced to obtain the optimum levels of cutting-tool and machining costs. Methods of calculating preset tool-life values which result in cutting conditions for optimum production rate and cost per unit have been explained by F. W. Taylor, W. W. Gilbert, J. Witthof, and others (1, 10, 17, 18).

Based on the mathematical relationship of production time per unit and the tool-life function, an equation for optimum tool-life time can be derived by differentiation as shown in the General Electric data system subsection of this chapter. That equation is:

$$T_{max.prod.} = \left(\frac{1}{n} - 1 \right) TCT \qquad (3\text{-}11)$$

The relationship for optimum tool-life cost can be found in the same way by using the unit cost and tool-life equation, and is the following:

$$T_{min.cost} = \left(\frac{1}{n} - 1 \right) \frac{t}{M} + TCT \qquad (3\text{-}9a)$$

After estimated labor and machine costs are introduced, tool-life preset values can be calculated. The tool-life preset values then take on the important function of transfer parameters resulting from considerations of work content, and they can be used as a basis for determining cutting values. Because a tool with specific cutting edges will usually be used for machining only a limited group of materials, optimized tool-life time and cost can be best met by the entry of the tool-life preset value, $T_{perm.}$, on the tool index card as shown in Fig. 3–45.

The EXAPT tool-life system permits the use of the same tool in the tool sets of different machines and yet enables full adaptation of operating conditions to each specific use. Optimum tool-life preset values for the same tool can then differ considerably on various cards, depending on the machine used.

Part Programming

The part program also contains input data for the EXAPT processors (Fig. 3–43). Spindle speed and feed values, rapid traverse commands, and all cutting commands are generated by the EXAPT processor and are transmitted as cutter location data (CLDATA) to the postprocessor (19).

Fig. 3–48 shows an engineering drawing of a shaft to be machined on an N/C lathe. The EXAPT 2 part program for this part, as shown in Fig. 3–49, is an example of how cutting values are determined. The program begins with heading statements

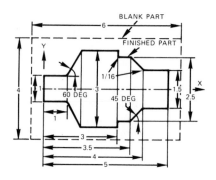

Fig. 3–48.—Engineering drawing of the lathe-machined shaft described in the text example.

```
                    ****   EXAPT 2 – PROCESSOR   ****

                    ** *   VERSION SO/8 OCT 69 SECTION 1   ****

                    ****   IMPLEMENTED ON UNIVAC 1108 ****

        1     PARTNO  / SHAFT 82470
        2     machin  / NCL
        3     UNITS   / INCH
        4     CLPRNT
        5     MACHDT  / 30 , 900, 0.004, 0.1, 5, 3600 , 1.0 , INCH
        6            CONTUR  / BLANCO $$ DESCRIPTION OF BLANK PART
        7            BEGIN   / -0.5 , 0 . YLARGE , PLAN , -0.5
        8            RGT     / DIA , 4
        9            RGT     / PLAN, 5.5
       10            RGT     / DIA , 0
       11     TERMCO
       12            SURFIN  / FIN      $$ PART SURFACE SHALL BE FINISHED
       13            CONTUR  / PARTCO   $$ DESCRIPTION OF FINISHED PART
       14     A,     BEGIN   / 0, 0, YLARGE, PLAN ,0, ROUGH
       15     B,     RGT     / DIA, 1 , ROUND, (1/16)
       16            LFT     / (LINE//POINT/ 1, 0.5), ATANGL, 60)
       17     C,     RGT     / DIA, 3
       18     D,     RGT     / PLAN 3 , ROUND , (1/16)
       19            LFT     / DIA , 2.5
       20            RGT     / (LINE//POINT/ 3.5, 1.25), ATANGL, 45), ROUND, (1/16)
       21            LFT     / DIA , 1.5
       22     E,     RGT     / PLAN , 5 , ROUGH
       23     F,     RGT     / DIA , 0
       24            TERMCO
       25     PART   / MATERL , 2300
       26     CSRAT  / 60
       27     CLDIST  / (1/16)
       28            FACING = TURN / SO , CROSS, TOOL, 101.1, SETANG, 180, ROUGH
       29            RGHCUT= TURN / SO , LONG , TOOL , 111.2, SETANG, -90, ROUGH
       30            FINCUT = CONT / SO , TOOL, 111.2, SETANG, -90, FIN
       31            CHUCK   / 50, 0 , 8 , 1 , 1 , -2
       32            CLAMP   / -0.5        $$ FIRST CLAMPING
       33            CUTLOC  / BEHIND      $$ CUTTING BEHIND THE AXIS
       34            WORK    / FACING
       35            CUT     / E , TO, F
       36            WORK    / RGHCUT , FINCUT
       37            CUT     / E , RE, C
       38            CLAMP   / 3, INVERS  $$ SECOND CLAMPING – INVERS
       39            WORK    / FACING
       40            CUT     / B , RE, A
       41            WORK    / RGHCUT , FINCUT
       42            CUT     / B, TO , D
       43     WORK / NOMORE
       44            FINI
```

Fig. 3–49.—EXAPT 2 part program for the example shaft.

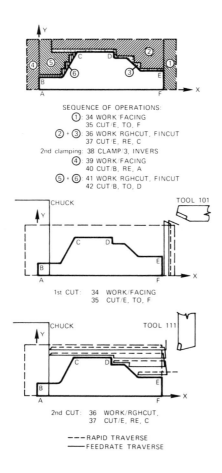

Fig. 3–50.—Sequence of operations and cuts
for the example shaft.

and the machine postprocessor selected in line 2. Line 5 gives data on the machine tool, including power (30kW), torque (900 ft-lbs), smallest (.004 ipr) and largest (.1 ipr) feed rates, and smallest (5 rpm) and largest (3600 rpm) spindle speeds. This information is used to determine cutting values. Lines 6 through 11 contain the description of the part blank, and lines 13 through 24 contain the description of the finished part. Line 12 indicates that the surface of the part is to be finished unless otherwise defined in the contour description. For example, in lines 14 and 22 the required surface quality for certain areas of the part is given as ROUGH.

Part material is given in line 25 of Fig. 3–49. This statement directs data for material number 2300 to be taken from the materials file. Statement 28 describes the machining operation for facing the front sections of the part. This is a rough-turning operation which uses tool 101 with a setting angle of 180 deg. Tool 101 is described in the tool file as shown in Fig. 3–45.

In statement 29, a rough-turning operation is defined, and in statement 30 the contouring operation to finish the part is defined. Line 31 describes the chuck to be used, and line 32 defines the clamping position. In statements 34 through 37, the right end of the part, as shown in Fig. 3–50, is machined. Turning operations are called up by WORK statements, and the sections of the part are described by CUT statements. CUT statements use the "contour marks" which were assigned to certain points of the part in statements 14, 15, 17, 18, 22, and 23. After inverse clamping, the left side of the part is machined.

Naturally, this example can show only some aspects of EXAPT 2 part programming. For flexibility, the parts programmer is given many possibilities for influencing machining and determining cutting values. For example, he can give correction values for spindle speed and feed rate in the PART statement, considering the stability of the part. Also, in order to obtain a temporary constant cutting speed for plane motions, he may enter a figure that represents the percentage of determined cutting speed below which the actual value should not fall. This percentage is stated as CSRAT (cutting-speed ratio), as shown in statement 26 of Fig. 3–49. When the machine tool is a lathe with stepped revolution rates, this statement results in a changed revolution rate when a plane is turned. Fig. 3–50 shows the operations which the part undergoes as a result of this part program.

Cutting Value Computation

After a machining location is described in the part program (statement 35 in Fig. 3–49, for example), the technological processor determines the area to be machined on the basis of a table compiled in the geometrical processor from blank and finished-component descriptions. This table contains all contour elements in a uniform numerical form. If examination shows that the contour cannot be obtained with the requested operation, then the operation is revised by the technological processor in such a way as to remove as much excess material from the blank as possible. Collision calculations are made which include (1) feed direction, (2) angle of approach, (3) main cutting-edge and side-rake location, (4) corner radius, and (5) tool shank and chuck dimensions. The area to be machined is divided into individual cuts in a cutter clearance program. Feed rates and cutting speeds are recalculated for each change in tool path due to changed conditions at the cutting-edge contact point.

The procedure by which cutting conditions are determined is shown in simplified form in the flow chart of Fig. 3–51. For simplification, the chart assumes a longitudinal turn, although the program provides for different factors that influence chip formation during simultaneous overlapping of plane movements or recessing operations. Data from the index cards is available to the core memory through the machine name, the material number, and the tool number called up from the edited part program. Information is available only for those cutting-tool materials that are permitted for machining the part material. The processor can then easily determine whether the tool can be used. If the tool cannot be used, an error message is printed out.

A value as large as possible is given for the chip width, b, in order to reduce the number of cutting passes. The value of b is limited either by the depth of cut programmed in the machining definition, by the maximum length of the cutting edge of the tool, $b_{perm.}$, or by the amount of material to be cut off. The machinable area is subdivided into single passes by the smaller value. A minimum depth of cut based on the data contained in the material and tool files is, as far as possible, accepted as the lowest limit in the division of cuts. Calculation of the tool path provides a table of coordinates for each cut. The feed rate is calculated first, followed by the cutting speed, for the target points of the work movements in this table.

For roughing operations, an initial feed-rate value is obtained from the machining width, b, and the setting angle. This method of derivation enables the user—by supplying a few reference values—to specify a desired ratio for width to thickness of the chip and to alter this function for cutting-edge geometry changes by a factor contained in the tool file. In this way, critical chip cross-sectional shapes can be controlled for every material/cutting-tool-material combination. For example, chips of oversquare cross-sectional shape can be avoided in shallow roughing cuts. Fig. 3–52 shows the theoretical hyperbola describing this mathematical relationship and the actual curve as it is modified by a practical example.

This parameter is not considered when determining finishing feed rates, because the small machining widths involved may result in lower feed rates than are required for surface finish quality. The part elements described in the part program as finishing or fine-finishing areas prescribe the maximum roughness and establish the feed rate for finishing cuts.

Cutting speed is then calculated from depth of cut and feed rate, considering tool life and the width of the flank wear land of the tool. The tool-life equation is only applicable to that range of cutting speeds for which approximate wear behavior is assured. The limiting values $V_{min.}$ and $V_{max.}$ are incorporated into the material files and are followed by the program. If the constant and exponents are selected correctly, a cutting speed can be determined which, together with depth of cut and feed rate, fulfills the conditions of wear and tool life. Fig. 3–53 shows the theoretical curve created by the tool-life/cutting-speed equation and the actual curve which is a result of all the limitations of machine tool, cutting tool, and materials.

EXAPT Management

The separation of input data for EXAPT processors into a part program and technological data files saves computer time for analysis, interpretation and compilation of cutting data. Risk of error is decreased because cutting values need not be given anew for each part program. Programming effort is considerably reduced, and the working conditions no longer depend on the part programmer's personal experience and diligence. Experience that has been found useful in production can be made part of the system, thereby ensuring that the company's knowledge is maintained despite a lack of qualified staff or changes in personnel.

The contents of the technological index cards can be prepared, edited, and kept up to date, independent of the actual parts programming work, by a metal-cutting expert. The file custodian gathers the experience gained in the company's parts programming and production departments, the recommendations of tool and materials manufacturers, and new findings in metal-cutting research, and makes this information available to the company without delay by simply updating the data files.

Several service programs are already available for updating purposes. The EXAPT 2 materials index, for example, can be mechanically compiled, documented, and tested independent of the processor, as shown in Fig. 3–54. This testing is accomplished by entering machining reference values from production and field tests in the program, which then prepares the corresponding materials index cards with all the necessary parameters. Another program simulates, by means of the tool and machine index cards,

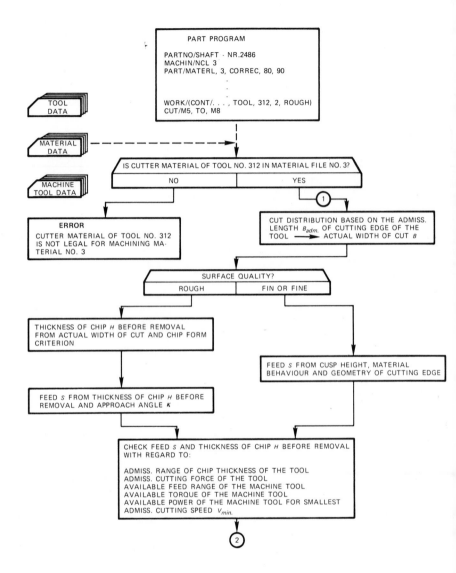

Fig. 3–51.—Flow chart showing calculation of depth of cut, feeds, and speeds in EXAPT 2.

Fig. 3–51.—(*Continued*)

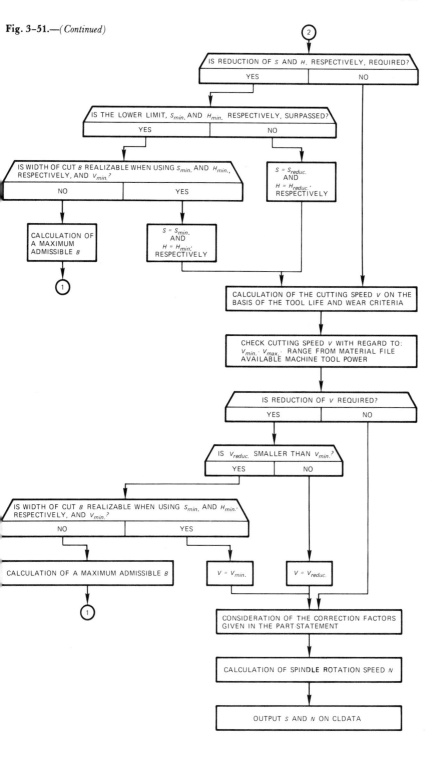

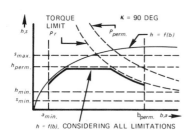

h = f(b), CONSIDERING ALL LIMITATIONS

Fig. 3–52.—Graph showing determination of feeds in EXAPT 2.

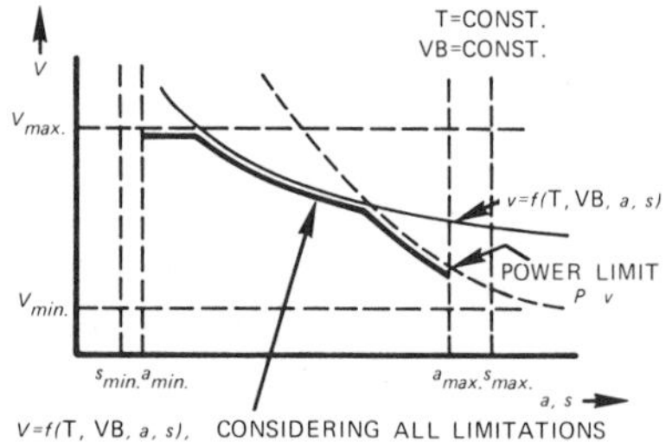

v=f(T, VB, a, s), CONSIDERING ALL LIMITATIONS

Fig. 3–53.—Graph showing determination of speeds in EXAPT 2.

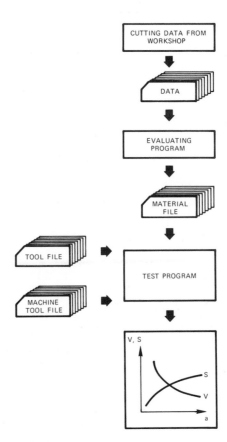

Fig. 3–54.—Technological evaluation program for creating and testing material data files for EXAPT.

the determination of cutting values by the technological processor and graphically plots the result on the high-speed printer.

Figs. 3–55 and 3–56 show an example of the output of this test program for the material data file, the machine-tool data file, and the cutting-tool data file (tool #101, see Fig. 3–45) used in the part program described in Fig. 3–49. The assumed approach angle, κ, is 70 deg, the same angle that is used when the tool faces the front planes of the part. The lower section of Fig. 3–55 shows the table resulting from the calculation of feed rate and cutting speed for individual depths of cut.

In addition, the input data on machine, tool, and material are printed out. Fig. 3–56 shows a graphical plot of the table in Fig. 3–55. This plot includes the curves for feed rate and cutting speed, both plotted over the depths of cut. The theory of both curves has been explained before in Figs. 3–52 and 3–53. With this diagram it is easy to find the feed rate and cutting speed for a certain depth of cut. For example, a depth of cut of .2 in. would require a cutting speed of 210 sfpm and a feed rate of .02 ipr as indicated by Fig. 3–56. The diagram covers the entire range of cutting depths permissible for a specific tool. With this system a firm can centrally handle machining data for the entire range of metal-cutting production and keep this data up to date with the help of a very small staff.

When EXAPT is compared to APT, companies using the EXAPT 2 system report a decrease in tape preparation costs due, for the most part, to a considerably shorter

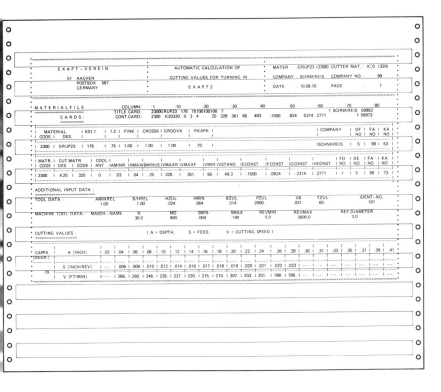

Fig. 3–55.—Printout of automatic calculation of cutting values in EXAPT 2.

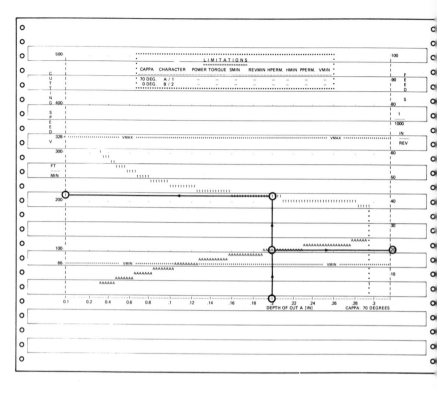

Fig. 3–56.—EXAPT 2 computer graph printout of automatically calculated cutting values.

part programming time. Furthermore, for many parts the actual machining time was much shorter with EXAPT 2 programming than with manual programming or APT programming because of the optimization of cutter-path and machinability data.

ABEX SYSTEM*

The Abex data system incorporates many historical developments in the field of metal removal and has many points in common with the other computerized systems described in this chapter. It, too, is a method of determining useful metal-cutting parameters for N/C and conventional machine tools by use of a high-speed digital computer. Also, before the Abex system can be employed, a performable process sequence and a part program must be developed. Several different types of N/C programs can be used, and most of them contain many of the essential elements needed to provide input data for this system. In addition, the system contains provisions for a continuous upgrading process which may be performed by several levels of operating personnel.

---

*The Abex computerized machinability data system is a proprietary system of the Abex Corporation. Requests for additional information concerning the system and its use should be addressed to the Abex Corporation, Manufacturing Research and Development, Mahwah, New Jersey.

General System

A part of the mathematical computer model used in the Abex system is derived from Taylor's expression for cutting speed, $VT^n = C$ (Eq. 3–2)(1). However, the limits of workable cutting conditions must be given careful consideration when minimum-cost and maximum-production-rate calculations are made using this equation. When these limits are not properly taken into account, the results may often be relatively low speeds, high feeds, and, in certain cases, heavy depths of cut which actually cannot be performed. When full-capacity, minimum-cost cutting values are suddenly introduced into production, exceeding these limits may result in inaccuracies in predicted cutting performance and other restraints which limit metal removal before the desired cutting conditions can be attained. Therefore, the final result of a single improper value in the part program may often be a totally unattainable standard or many scrap parts. This problem can be solved by the addition of adjustment factors to the mathematical model and the use of decision tables. Limiting values are also developed and used in the proprietary machinability programs prepared by carbide tool manufacturers.

The Abex system contains both a synthesis of previous solutions and new techniques to solve machining problems within cutting-condition limitations. It makes use of the complex manipulations of commercial work-measurement computer programs which include methods of (1) describing a part in a form similar to that of N/C techniques, (2) storing feeds, speeds, horsepower, and machine-tool parameters, and (3) generating cutting conditions on short-cycle operations. The cost calculations used are based on W. W. Gilbert's concepts of the four basic contributions to cost and on computer programs developed by Metcut Research Associates, Inc., which detail alternative machining parameters (10, 20, 21, 22, 23). A more detailed discussion of these cost calculations is contained in Chapter 4.

In the Abex system, shop-generated machining data are organized and selected on the basis of "operation family" concepts, and tool life is defined on the basis of the tool's ability to hold a specified tolerance or surface finish. The term *operation family* is used to mean certain operation relationships which for a given plant have been shown to have an important influence on the accurate and precise performance of calculated process parameters. For example, in turning, workpiece diameter is often one such parameter, and all operations involved in obtaining a given range of diameters make up the operation family. In addition, all long-cycle operation parameters are determined on the basis of economic considerations. The system's experimental data bank is searched to find an additional set of operating parameters based on the minimum-cost feed rate selected from the shop-generated data.

The Abex system provides for the selection of metal-cutting conditions based on strategies other than minimum cost or maximum production. The printout of cutting conditions contains sufficient data to allow the parts programmer to specify tooling and cutting parameters that will reproduce those recommended conditions. If the conditions cannot be attained on test workpieces, other machine-tool settings can be found systematically based on cost considerations.

Data Storage Section

It is well within the range of available computers to store machinability data on the most commonly used workpiece materials. This storage capability eliminates errors that result from the conversion to exponent values and reconversion to cutting-condition data for each operation. One approach to using this computer storage capability is to allow results of successful, adequately detailed shop studies to be added to the computer's stores of useful machining information. By this method, the results of particularly successful high material removal rates can be recalled at any time.

The operation of the basic Abex program is shown in the general system flow chart of Fig. 3–57. The data storage section of the system consists of the following five data files:

1) Machine-tool data (DB 100)
2) Base cutting-condition data (DB 200)
3) Experimental data (DB 300)
4) Shop-generated data (DB 400)
5) Plant cost data (DB 500).

Machine-Tool Data (DB 100). The machine-tool file contains (1) machine identification and description, (2) available horsepower, (3) speeds, and (4) feeds for each machine. It also provides for analysis and simulation testing with hypothetical "machines" having infinitely variable feeds, speeds, and available horsepower. All descriptive data are entered into the computer random-access storage area and are updated when new machines are purchased. The descriptive data available from machine-tool builders contain additional information, such as rapid traverse times, programmed acceleration and deceleration cycles, and tool-change times, which can also be programmed and incorporated in the machine-tool data bank.

Base Cutting Conditions (DB 200). Base cutting conditions for specific types of materials are stored in this section. Briefly, this file contains (1) the material code, (2) unit horsepower per cubic inch, (3) cutting speed and feed for HSS and cemented carbide tools, and (4) a description for each material used.

Modifications to base cutting conditions are developed by the use of adjustment factors. For example, speed factors are based on the ratio of recommended speed for a particular operation (for the base material under base cutting conditions) to the base speed for the base operation (turning at 170 sfpm for HSS and at 600 sfpm for cemented carbide tools) (23).

Experimental Data (DB 300). The results of laboratory tests are contained in the experimental data bank. Each line of data contains the following information:

1) Operation type
2) Feed, speed, and depth of cut
3) Tool material
4) Tool geometry
5) Work material
6) Work material hardness
7) Average cutting force values (optional)
8) Surface finish imparted to the workpiece
9) Tool life as a function of one or more principle qualities imparted to the workpiece.

This last information, the tool-life factor, is of great importance. For example, when a given carbide grade of turning tool is used for a roughing operation, a common tool-life end point is a .025-in. flank wear land. In the case of finishing cuts, when a harder, brittler carbide is used, the ability to maintain a 125 microinch finish or size tolerance within a specific range is often the major factor. As another example, the tool life of a ball end mill is often a function of the tool's ability to maintain a given part size tolerance. In any case, the point here is that the workpiece material and its hard-

DATA STORAGE SECTION   OPERATIONAL SECTION

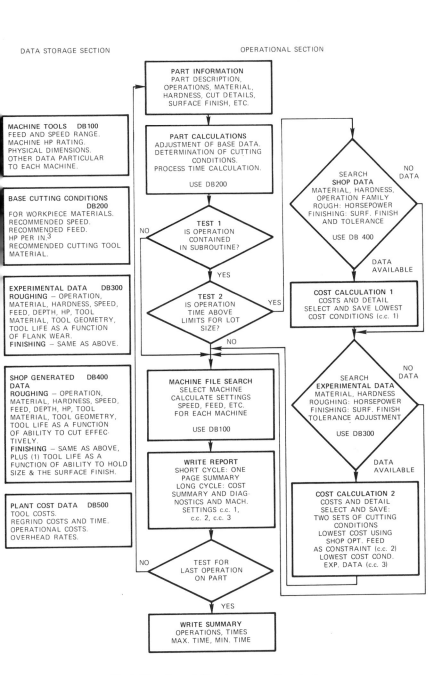

Fig. 3–57.—General Abex data system flow chart.

ness are associated with a specific tool material and geometry and that tool performance is associated with all these data.

In most plants, this experimental information is collected but is often not organized or recorded. It is relatively easy, however, to describe N/C tooling systematically. For example, a single-point tool geometry described by cutting angles of −5, −5, 5, 5, 15, and 15 deg and by a nose radius of 3/64 in. could be filed as B3. The B indicates the basic geometry, and the 3 denotes the nose radius. Special attention should be given to the nose radius because of the important relationship of this dimension with workpiece surface finish. In addition, for the convenience of shop personnel, toolholder identification can be directly related to this scheme. The program can then generate toolholder descriptions from the alphanumeric code. The main intent of this recording system is to present sufficient information to allow duplication of full-capacity optimum conditions or to allow adjustments based on cost considerations.

Shop-Generated Data (DB 400). Information contained in the shop-generated data file is the result of successful experience in the shop. Foremen, industrial engineers, and others can observe operating conditions and report the following data to be key-punched and entered into the file:

1) Operation type
2) Work material
3) Work material hardness
4) Speed, feed, and depth of cut
5) Horsepower (optional)
6) Tool material
7) Tool geometry (coded)
8) Operation family
9) Tool life.

This section, then, contains the best shop data available on successful operations. It provides additional benefits in that it contains appropriate details which can be recalled when similar conditions are encountered with a new workpiece. Machinability data for this file can be collected from conventional machine tools of the same type as their N/C counterparts, and very often this method is preferable because it allows the operator supervisor or engineer more freedom to vary cutting conditions and to observe the results during machining.

Plant Cost Data (DB 500). The plant cost data file contains the cost factors required to compute a full-capacity optimum cutting condition for each operation. The significance and determination of cost accounting values must be taken into consideration in order to allow the optimum to reflect accurate minimum-cost conditions.

Operational Section

The operational section of the Abex system is also shown in Fig. 3–57. In its operations the computer performs a series of tests, calculations, and data file searches and then prints out a summary of operations, times, and optimum cutting conditions.

Part Calculations. The first unit of the operational section is a part information input facility. This unit accepts the part description coded in a digital format and includes certain other machining factors such as depth of cut, surface finish required, and related requirements. The workpiece description is passed into several calculating subroutines which search for the appropriate base conditions, make adjustments, and generate process time calculations using DB 200 (23).

Tests 1 and 2. Next, two tests are performed. The first determines the presence of the appropriate subroutine for turning, milling, etc. The second test establishes the time of a specific operation in relation to annual production. For example, if the required number of parts is twenty, the time of an operation to warrant further consideration might be three or four minutes. This is termed a long-cycle operation. If the first or both tests are negative, then the program continues without entering this subroutine section.

Shop Data Search. When a subroutine is entered, however, the subroutine section first considers data generated in the shop (DB 400). This data is processed on the basis of a series of qualifications. The entire bank of data is searched to find similarities between the part being processed in the program and the historical data stored in the data bank. The order of search is by (1) basic workpiece material, (2) workpiece hardness within a percentage range, and (3) operation family.

Cost Calculation 1. In the next step, cost calculations are performed, and the lowest cost conditions are selected. Figs. 3–58, 3–59, and 3–60 illustrate detailed analyses for long-cycle operations in turning, drilling, and end milling, respectively. The first part of the computer output describes the cutting conditions retrieved from the shop-generated data bank, DB 400. In addition, the tool cost and tip cost are calculated to find dollars per hundred pieces produced. These costs represent the contribution of expendable tool cost to total cost. The column headed MACH COST (machine cost) represents the overhead rate on the machine tool multiplied by the cutting time for the operation.

The total cost and production rate are also based on a per-hundred-units-produced standard. Total cost is the sum of overhead rates multiplied by the time to perform the following functions during the specific operation:

1) Rapid traverse
2) Feed
3) Load and unload
4) Setup

MACHINABILITY ANALYSIS – COST SUMMARY AND CUTTING CONDITION DETAILS

PART NO. 03189–T OPER NO. 100 DATE 03/69
MATERIAL DESCRIPTION – 4340B 0.86 HP/CU. IN.
OPERATION NAME – TURN

CUT DETAILS – DIAMETER – 3.5000 LENGTH – 19.0000 DEPTH – 0.100
FINISH – 125. RMS

CALCULATIONS USING SHOP DATA

CUT SPEED RPM	CUT SPEED SFPM	FEED RATE IPR	FEED RATE IPM	TOOL LIFE MIN.	WORK DIAM IN.	TOOL DESCRIPTION OP****TH****RAK **MATL**NR***LA	SIZE TOLER IN.	SURF FIN URMS	TOOL COST $/100	TIP COST $/100	MACH COST $/100	TOTAL COST $/100	PROD RATE HRS/100	PARTS PER T.L.
448.	470.	0.010	4.48	15.	4.000	11 78 0103P0600	0.005	125.	0.19	5.35	63.54	125.61	9.15	3.54
381.	400.	0.010	3.81	30.	4.000	11 78 0103P0600	0.005	125.	0.11	3.14	74.66	128.33	9.71	6.02
343.	360.	0.010	3.43	45.	4.000	11 78 0103P0600	0.005	125.	0.08	2.32	82.96	133.52	10.38	8.13
310.	325.	0.010	3.10	60.	4.000	11 78 0103P0600	0.005	125.	0.07	1.93	91.90	140.96	11.25	9.79

MINIMUM COST CONDITIONS, SHOP DATA

512.	470.	0.010	5.12	15.	3.500	11 78 0103P0600	0.005	125.	0.19	5.35	63.54	125.61	9.15	3.54

MACH OVHD $/MIN	NO. CUTS	APPR TO WRK IN.	RPDTRV COST $/100	LOAD TIME MIN.	LOAD COST $/100	SETUP TIME MIN.	SETUP COST $/100
0.15	1.	4.0	4.05	2.3	34.49	21.0	3.14

MACHINE GROUPS – GROUP NO. MACHINE DESCRIPTION MACH HP
1647 HORZ. TURRET LATHE W&S UNIV. 3 5 HP 3.500

1 RPM– 352. FPR– 0.0075 IPM– 2.63 HRS/100– 12.84
2 RPM– 460. FPR– 0.0075 IPM– 3.44 HRS/100– 9.93
3 RPM– 564. FPR– 0.0075 IPM– 4.22 HRS/100– 8.18

Fig. 3–58.—Detailed Abex machinability analysis for a turning operation.

MACHINABILITY ANALYSIS – COST SUMMARY AND CUTTING CONDITION DETAILS

PART NO. 03189–T OPER NO. 200 DATE 03/69

MATERIAL DESCRIPTION – 4340B 0.86 HP/CU. IN.

OPERATION NAME – DRILL

CUT DETAILS – DIAMETER – 0.2500 LENGTH – 0.5000 DEPTH – 0.125

FINISH – 125. RMS

CALCULATIONS USING SHOP DATA

CUT SPEED RPM	CUT SPEED SFPM	FEED RATE IPR	FEED RATE IPM	TOOL LIFE MIN.	TOOL DIAM IN.	TOOL DESCRIPTION OP****TH****LNTH **MATL** ANG ***	SIZE TOLER IN.	SURF FIN URMS	TOOL COST $/100	SHARP COST $/100	MACH COST $/100	TOTAL COST $/100	PROD RATE HRS/100	PARTS PER T.L.
1374.	90.	0.002	2.74	3.	0.250	21HSS102118JOBL	0.005	125.	3.40	20.84	13.64	63.41	2.26	3.
1129.	74.	0.002	2.26	11.	0.250	21HSS102118JOBL	0.005	125.	1.35	8.29	16.59	50.62	2.46	9.
916.	60.	0.002	1.83	27.	0.250	21HSS102118JOBL	0.005	125.	0.67	4.12	20.46	49.26	2.85	20.
763.	50.	0.002	1.52	52.	0.250	21HSS102118JOBL	0.005	100.	0.41	2.56	24.56	51.40	3.29	32.
1129.	74.	0.005	5.64	0.	0.250	21HSS102118JOBL	0.005	125.	6.61	40.56	6.63	81.13	1.68	2.
687.	45.	0.005	3.43	7.	0.250	21HSS102118JOBL	0.005	125.	1.34	8.22	10.91	44.85	1.83	10.
549.	36.	0.005	2.74	18.	0.250	21HSS102118JOBL	0.005	125.	0.67	4.12	13.64	42.44	2.09	20.
458.	30.	0.005	2.29	39.	0.250	21HSS102118JOBL	0.005	125.	0.37	2.29	16.37	42.87	2.38	36.

MINIMUM COST CONDITIONS, SHOP DATA

549.	36.	0.005	2.74	18.	0.250	21HSS102118JOBL	0.005	125.	0.67	4.12	13.64	42.44	2.09	20.

MACH OVHD. $/MIN	NO. CUTS	APPR TO WRK IN.	RPDTRV COST $/100	LOAD TIME MIN.	LOAD COST $/100	SETUP TIME MIN.	SETUP COST $/100
0.15	5.	3.0	4.87	1.0	14.99	15.0	3.74

MACHINE GROUPS – GROUP NO. MACHINE DESCRIPTION MACH HP

0824 DRILL PRESS L & G 2 SPINDLES 1/4 HP 0.180

1 RPM– 150. FPR– 0.0050 IPM– 0.74 HRS/100– 6.24
2 RPM– 150. FPR– 0.0050 IPM– 0.74 HRS/100– 6.24
3 RPM– 225. FPR– 0.0050 IPM– 1.12 HRS/100– 4.34

Fig. 3–59.—Detailed Abex machinability analysis for a drilling operation.

5) Tool change
6) Tool depreciation
7) Tool resharpening, rebrazing, or blade reset
8) Insert or blade cost and grinding wheel cost.

Only the machine cost, tool costs, and tip costs (costs associated with tool wear, such as resharpening cost and blade cost) are printed for each of the cutting conditions. The

MACHINABILITY ANALYSIS – COST SUMMARY AND CUTTING CONDITION DETAILS

PART NO. 03189–T OPER NO. 300 DATE 03/69

MATERIAL DESCRIPTION – 4340B 0.86 HP/CU. IN.

OPERATION NAME – END MILL

CUT DETAILS – DIAMETER – 0.7500 LENGTH – 8.0000 DEPTH – 0.750

FINISH – 125. RMS

CALCULATIONS USING SHOP DATA

CUT SPEED RPM	CUT SPEED SFPM	FEED RATE IPM	FEED RATE IPR	TOOL LIFE MIN.	TOOL DIAM IN.	TOOL DESCRIPTION OP****TH****ANG **MATL**RAKE***	SIZE TOLER IN.	SURF FIN URMS	TOOL COST $/100	SHARP COST $/100	MACH COST $/100	TOTAL COST $/100	PROD RATE HRS/100	PARTS PER T.L.
351.	69.	1.60	0.004	6.	0.750	31HSS104PBPB45	0.0050	125.	52.92	148.18	83.43	397.53	16.32	1.11
290.	57.	1.20	0.004	21.	0.750	31HSS104PBPB45	0.0050	125.	20.06	56.19	111.25	262.94	15.24	2.93
239.	47.	1.06	0.004	51.	0.750	31HSS104PBPB45	0.0050	125.	9.71	27.19	125.94	226.44	15.56	6.07
193.	38.	0.90	0.004	35.	0.750	31HSS104PBPB45	0.0050	125.	16.66	46.66	148.33	283.21	18.93	3.53
422.	83.	1.80	0.004	16.	0.750	31HSS204PBPB45	0.0050	125.	17.57	49.19	74.16	213.51	10.80	3.35
351.	69.	1.60	0.004	31.	0.750	31HSS204PBPB45	0.0050	125.	10.51	29.44	83.43	187.92	10.93	5.60
279.	55.	1.20	0.004	57.	0.750	31HSS204PBPB45	0.0050	125.	7.62	21.34	111.25	201.42	13.66	7.73
193.	39.	0.90	0.004	40.	0.750	31HSS204PBPB45	0.0050	125.	14.58	40.83	148.33	272.91	18.66	4.04

MINIMUM COST CONDITIONS, SHOP DATA

351.	69.	1.60	0.004	31.	0.750	31HSS204PBPB45	0.0050	125.	10.51	29.44	83.43	187.92	10.93	5.60

MACH OVHD $/MI	NO. CUTS	APPR TO WRK IN.	RPDTRV COST $/100	LOAD TIME MIN.	LOAD COST $/100	SETUP TIME MIN.	SETUP COST $/100
0.15	1.	6.0	2.99	3.0	45.00	30.0	4.50

MACHINE GROUPS – GROUP NO. MACHINE DESCRIPTION MACH HP

1942 MILLER HORZ. CINN. 2MI 5 HP 3.500

1 RPM– 295. FPR– 0.0045 IPM– 1.25 HRS/100– 12.35
2 RPM– 385. FPR– 0.0045 IPM– 1.75 HRS/100– 8.85
3 RPM– 500. FPR– 0.0045 IPM– 2.50 HRS/100– 6.23

Fig. 3–60.—Detailed Abex machinability analysis for an end-milling operation.

remainder of the factors are included in the total cost. Those factors that are constant for an operation are printed below the section entitled CALCULATIONS USING SHOP DATA. In addition, other operation details are printed on this line. The machine-tool overhead rate is presented. Machine overhead can be a variable input, and changes in its value, while the values of other variable costs are maintained, allow sensitivity analyses to be made. Thus, by lowering overhead rates as a function of capacity demands on a machining center, the accounting-based overhead contributions may be reduced, and the expendable tool-cost factor may be raised.

Experimental Data Search. The search of experimental data in the next step of the system is similar in nature to the shop-generated data processing just discussed. The major difference between the two is that tolerance must be made a function of flank land wear in the experimental data search. Use of the flank wear land is probably the most accurate method, but further experience and experimentation are required in this area.

Cost Calculation 2. The calculation section for experimental data computes a total cost per piece for each set of cutting conditions. From this group, two sets of cutting conditions are selected. The first is the least-cost condition at a feed less than or equal to the minimum-cost feed for shop-generated data. Thus, some of the complex and undefined factors such as rigidity are taken into account. The second set of cutting conditions selected is that for absolute minimum cost for a full-capacity optimum. These three sets of conditions—(1) the optimum conditions from the shop-generated data, (2) the feed-constrained experimental data conditions, and (3) the minimum-cost conditions from experimental data—are introduced into the machine file search section.

Machine File Search. The input machine groups are searched for the speed and feed closest to the calculated values. Each machine group is searched for a rotational rate that is within a predetermined percentage value of the calculated rate. The machine data group is interrogated for the feed closest to that calculated.

A final and important step in the ultimate determination of viable and economical cutting conditions is the matching of determined cutting conditions to the feed and speed settings available on the machine tool. This is a complicated procedure to perform manually, but is readily performed by data-processing equipment.

Write Report. Finally, as shown in Fig. 3-57 in the WRITE REPORT operation, a document is prepared. This document can be in a short form, or, if the subroutine is activated on the basis of the predescribed conditions, it will contain the descriptions shown in Figs. 3-58 through 3-60.

Computer Operations

Many plants produce components which are similar in shape and function. The intent of the shop-generated data bank is to develop data within the computer system which is highly accurate and "knowledgeable" of all these products. This is a gradual process, but the capacity of the digital computer to store and recall large amounts of data in an orderly manner is an important element in its organization. This method is superior to many other computer systems which essentially "start from scratch" on each calculation and do not improve with experience.

Since development of the logic modules used to analyze specific operations within the search block is an important part of the system, a detailed illustration is presented below. The adaptation of a systems model to determine machining conditions requires the cooperation of both operating personnel and those persons involved with developing the generalized systems model. In other words, a framework can be developed remote from the manufacturing floor, but "tailoring" the model to a specific plant requires the knowledge and experience of people closely associated with plant operation—foremen, process planners, and manufacturing engineers. This point can best be illustrated by the following example.

Example—Internal Boring. Consider a process, such as an internal boring operation, commonly performed on parts manufactured in a particular plant. The framework of the general Abex system provides for the input of the operation name (internal bore), the dimensional parameters of the part, and other detailed information. For a turning operation, for example, this information is workpiece diameter, wall thickness, surface finish, tolerance, and other data. Each type of operation requires a pattern of logic within the search block in order to generate viable cutting conditions. To develop the logic for a specific operation, parameters must be listed, and cutting conditions must be sorted.

List Parameters. When the important parameters of each metal-cutting operation are listed, opinion must be separated from observed shop performance. Often a foreman, a process planner, and a manufacturing engineer can develop a list of these parameters. For internal boring, satisfactory cutting conditions may be found to be dependent on boring bar diameter, tool geometry, and several other parameters. These parameters can be listed as dependent and independent variables as follows:

Independent Variables	*Dependent Variables*
Feed	Size tolerance
Speed	Surface finish
Depth of cut	
Bar diameter	
Tool geometry	

In this approach, tables and reference sheets currently in use by process planners can be used. In addition, however, actual shop data which show the correlation between predicted and actual results are required.

Sort Cutting Conditions. The second step in the development of logic to control special operations is the sorting process by which the final cutting conditions are determined. This process is often directly patterned after the process planner's best approach to the problem. However, since the computer cannot use any intuitive judgment, as the process planner can, all the steps in the process must be listed. A list of these steps, in the form of questions to be answered, might be prepared as follows:

1) Is the operation an internal boring operation?
2) What are the dimensions of the bore?
3) What tolerance is required?
4) What surface finish is required?
5) What additional product specifications must be attained?
6) What speeds and feeds will give this result?
7) What is the most economical combination of feeds, speeds, and other factors to use in the job?
8) What additional parameters are pertinent to this operation?

Flow charts and FORTRAN programs are required to transform the logic pattern into a format acceptable to a computer. For the internal boring problem, an abbreviated flow chart might assume the form shown in Fig. 3–61. A sample of the FOR-

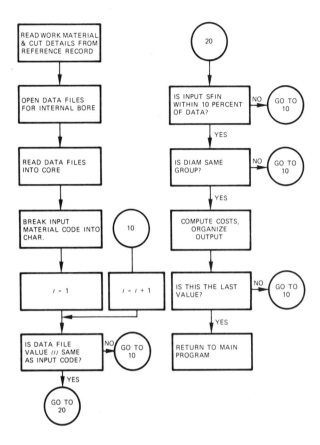

Fig. 3-61.—Abbreviated flow chart showing detailed analysis which takes place in the SEARCH and COMPUTATION blocks of the general Abex system flow chart, Fig. 3–57.

TRAN programming used to actuate the internal boring operation is shown in Fig. 3–62.

Logic Accuracy. As the machining operation is initially analyzed for the logic patterns that are needed to generate machining parameters, the analyst should realize that the logic will only be accurate as he chooses to make it. If the operation has been a typically difficult one to perform, and if many factors are required, then the logic developed to parallel the manual method of selecting cutting conditions will reflect this complexity. However, once the initial task is completed, the job becomes routine and a consistent answer can always be expected. Furthermore, as additional information is obtained about the operation or the logic used in selecting feeds, speeds, etc., then additions or deletions can readily be made to the data bank or the logic sequence itself.

Among the important considerations in logic development that should be emphasized are the following:

```
**              Programming—Internal Bore
                Dimension Statements
                Common LDR, MAN, NOMAN
                Equivalence Statements
C               Initialize File Read Displacement
                JDIS = 0
C               Read Work Mat'l. Description from Record
                Call FDATA (85)
                Call RDATA (85, 0, 2, AMAT (1))
C               Open Internal Bore Data File (100)
                Call FDATA (100)
                Do 100 I = 1, 50
C               Read A Line of Data Into Core
                Call RDATA (100, JDIS, 14, A(1))
C               Increment Read Displacement
                JDIS = JDIS + 14
C               Check For Last Data Item
                If (A(1)) 110, 110, 5
        5       Continue
C               Compare Material Codes
                If (A(13) − AMAT(1)) 100, 10, 100
        10      If (A(14) − AMAT(2)) 100, 20, 100
C               Material Code Matches—Compare Surface Finish
        20      If (SFIN * 1.10 − A(12)) 100, 30, 30
C               Surface Finish Is Less Than 10% Greater Than Input
C               Check Diameter For + Or − 10%
        30      If (DIAM * 1.10 − A(6)) 100, 40, 40
C               Diameter Is Less Than 10% Greater, Check Low Side
        40      If (DIAM * .90 − A(6)) 50, 50, 100
C               Diameter Is Good—Compute Costs
C               Compute Feed In Inches/Minute
        50      FIPM = A(1) * A(3)
C               Compute Tool Change Coefficient
                Alpha + DIAM * XLN/(3.82 * A(3) * A(2) * A(5)) * CPC
C               Compute Tool Change Cost/100 PCS.
                TLCHG = ALPHA * TC * XM * 100
                                        (TC  = Tool Change Time)
                                        (XM = Overhead)
C               Compute Tool Depreciation  (CP  = Tool Holder Cost)
                                        (XK1 = Holder Life Changes)
C               Compute TIP Costs
                TIP = ALPHA * CC/XK3 * 100  (CC  = Tip Cost)
                                        (XK3 = Cutting Edges)
C               Compute Machine Costs
                Feed=XLN*CPC/FIPM*100.*XM*(TRVDT+TRVDF)/RTR*
                1 XM * 100
C               Compute Total Cost For Machining (Not Including Set Up
C               Or Loading)
C               Compute Production Rate PC/HR
                PCPHR = 60./((FEED + TLCHG)/(XM * 100.))
C               Print Output
                Write (3, 60) (A(I), J = 1, 3), FIPM, (A(J), J = 5, 12), TDEPR,
                TIP, FEED, COST, PCPHR
        60      Format (Use Desired Spacing)
        100     Continue
        110     Call Exit
                Return to Main Program
```

Fig. 3-62.—Sample FORTRAN programming to perform the elementary tasks described in the flow chart shown in Fig. 3-61.

1) If the process planner or foreman has a rational, orderly system to apply, it can be readily transformed into a flow chart and a computer program.
2) The results of such an analysis will be consistent and permit a standardization of this phase of the process planning function.
3) New information (about materials, tools, etc.) can be readily added to the body of knowledge surrounding the performance of a machining operation.
4) The speed at which modern computers can manipulate data and apply a logically developed analysis to the selection of cutting conditions permits the consideration of a great many factors that traditionally would have been ignored or averaged.

ESTABLISHING A COMPUTERIZED SYSTEM

The relative importance of the different factors involved in implementing a computer-based machinability data system varies from plant to plant. Two of these factors may be (1) the availability of a computer and its acceptance as a manufacturing management tool, and (2) the technical competence of the process planners, parts programmers, and industrial engineers who will contribute to the system's establishment. Because of these and other variable factors, any plan to implement a computer-based system must be considered carefully.

Parts Analysis

The scope of the parts to be analyzed by the system must be defined. Often, an A-B-C ranking in order of part use and complexity is a useful tool. At the very least, such a definition will develop a ranking of parts by their machining costs. By introducing current yearly parts quantities and future projections, total dollar values can be calculated. In addition, detailed projections of savings can be made by analyzing selected typical parts from similar part family groupings.

Project Planning

A schedule detailing tasks and responsibilities should be made. This part of the planning process will probably involve the talents of several specialists. If the "milepost" technique for project documentation is used, the schedule may be arranged like that shown in Fig. 3–63.

Management Training

The full cooperation and acceptance of the system must be obtained from all personnel involved. To create this cooperation and acceptance, the exact nature of the program must be understood fully. Exposure of managerial and supervisory personnel to condensed segments of a training program is a workable method of developing this understanding of the system.

Fund Allocation

The exact allocation of dollars toward implementation of the system must be determined. To make the exact nature of the allocation decision clearer, it can be simplified. For example, the coding of part information to be entered into the system is essentially a variable cost. After personnel have been trained, and their learning-curve transition period has been passed, the cost of preparing input information becomes a direct linear function of the number of operations to be performed on the programmed parts. How-

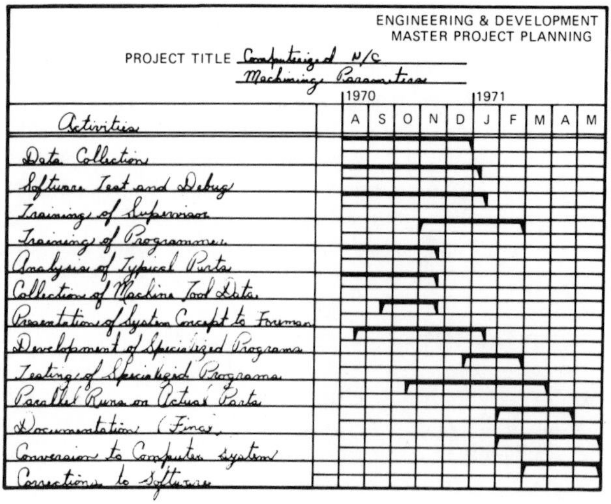

Fig. 3-63.—"Milepost" project planning schedule for establishment of a computerized machinability data system.

ever, the fixed cost of developing machinability data is added to the variable input data preparation cost. Basic information for machinability data development is available from outside sources, such as the Air Force Machinability Data Center, but some in-plant observation is still required to establish a frame of reference for data development costs.

In addition, programming costs for the logic section of the computer are involved. Complex well-conceived manipulative frameworks for the logic section are available from a number of computer manufacturers, but the experiences of several plants involved in developing computerized data systems indicate that it may be necessary to tailor these frameworks extensively to develop an accurate predictive model for the individual plant. This sort of tailoring is routine in developing many complex computer programs, but management must be accurately informed of the initial costs which will occur because of the data-processing skills required for such programming.

Parallel Operation

At first, the newly developed computerized data system should be operated in conjunction with previous techniques of machinability data development. Although this duplication of effort is disliked by some manufacturing managers, failure to satisfactorily "debug" the new system is a major cause of installation failure, and parallel operation is necessary until all data banks, analysis programs, and computer operations are operating smoothly. This experience will also give operating personnel added confidence in the precision and predictive accuracy of the system.

Personnel Training

The effective training of the personnel who will be using the new data system is a major step in attaining successful computer-oriented operations. Training for these people should emphasize the following:

1) Basic machinability concepts
2) Basic computer operation concepts
3) Operational experience.

Machinability Concepts. Basic concepts of machinability, such as the relationships between feed rate, chip cross-sectional area, tool deflection, and tool wear, should be presented to system operators and their supervisors. Often, parts programmers are presented only with machining condition tables. It is important that they also be exposed to the basic principles which underly the generation of such tables. They will accept the validity of a computer calculation easier if they realize the similarities between existing techniques and those used by the computer.

Computer Concepts. The basic concepts of high-speed digital computer operations should be discussed. Such a presentation should illustrate the similarities between the manual processes for determining economical cutting conditions and the processes aided by a computer.

Operational Experience. This final phase of the training program is the most important. It involves "hands-on" experience with the system—actually operating the computer to produce useful results. Sample parts which are representative of those to be processed in the plant should be selected and developed into case studies for the trainees to follow.

Machinability Data Improvement

In many situations, the individual parts programmer does not have detailed machining experience—certainly not an equivalent to that contained in a central computer data bank. In addition, he seldom has the time available to carry out the complex calculations required to select and adjust cutting conditions. Nor does the programmer have time to perform the economic calculations that are so easily produced by a computer. These economic calculations are especially important with the dramatic increase in overhead rate for N/C-produced workpieces.

Exponentially determined cutting values are often far removed from actual results. Unless elaborate constraints are used, workpiece spoilage or machine-tool damage is possible. When such constraints are applied, the likelihood of establishing overly conservative conditions is increased. Other problems also result. However, when adjustments to base cutting conditions are used, there still remains the potential for improvement; as the process approaches a theoretical optimum, it is necessary only to supply additional analysis and adjustment information to the process planner to allow increased efficiency.

In the machinability data models discussed in this chapter, attention has been given to the generation of usable manufacturing parameters. During the development of these systems, close liaison was maintained with operating personnel. In this way, the routine aspects of certain tasks were analyzed and incorporated in the systems for their improvement. Some of the benefits of this step are the following:

1) The systems are made self-improving and become refined with time. Constraints can be added, removed, or modified, and new tool-material developments, etc., can be readily incorporated.
2) Sensitivity analyses can be easily performed by entering, for example, overhead cost rates of $10, $15, and $20 and then examining their effect on minimum-cost cutting conditions. Also, comparisons of costs related to different tool configura-

tions (for example, disposable carbide inserts compared to brazed tool shapes) can be made.

3) The computers can be used to assist in increasing the productivity of the shop itself by performing a multitude of routine calculations on a stored data bank.

4) The computer-aided analyses described in this chapter can be operated by competent plant personnel. Currently employed industrial engineers can be upgraded by the emphasis on optimized operation and tool material selection.

5) Savings, as a result of special studies, may be documented for management reporting.

REFERENCES

1. Frederick W. Taylor, "On the Art of Cutting Metals," *Transactions of the ASME,* Vol. 28 (1907), No. 1119, pp. 31, 279.
2. Robert G. Brierley, "Cemented Oxides: Where, When and How to Apply," *SAE Paper 981B,* Jan., 1965.
3. *Manual on the Cutting of Metals with Single-Point Tools,* 2d ed., New York, American Society of Mechanical Engineers, 1952.
4. E. J. Weller and C. A. Reitz, "Optimizing Machinability Parameters with a Computer," *ASTME Paper No. MS66-179* (1966).
5. Joseph Datsko and O. W. Boston, "Relative Abrasiveness of the Cast Surfaces of Various Gray-Iron Castings on Single-Point Tools of High-Speed Steel," *Transactions of the ASME,* Vol. 75 (1953), p. 106.
6. *Selection and Application of Single-Point Metal-Cutting Tools: Publication GT9-270,* Detroit, General Electric Company, Metallurgical Products Department, 1969.
7. W. W. Gilbert and W. C. Truckenmiller, "Nomograph for Determining Tool Life and Power When Turning with Single-Point Tools," *Mechanical Engineering,* Vol. 65 (Dec., 1943), pp. 893–98.
8. Robert G. Brierley and H. J. Siekmann, *Machining Principles and Cost Control,* New York, McGraw-Hill Book Co., Inc., 1964.
9. *Operating Manual for the Carboloy Machinability Computer: Manual No. MC-101-B,* Detroit, General Electric Company, Metallurgical Products Department, 1957.
10. W. W. Gilbert, "Economics of Machining," in *Machining: Theory and Practice,* Metals Park, Ohio, American Society for Metals, 1950.
11. H. Opitz, et al., "Das Programmiersystem EXAPT (The EXAPT Programming System)," *TZ für praktische Metallbearbeitung,* Vol. 61, No. 6, pp. 404–14.
12. H. Opitz and B. Hirsch, "Programmation automatique des machines-outils à commande numérique (Automatically Programming N/C Machine Tools)," paper presented at the C.I.R.P. European Conference on the Numerical Control of Machine Tools, Paris, 1968.
13. Bernd Hirsch, "Bestimmung optimaler Schnittbedingungen bei der maschinellen Programmierung von NC-Drehmaschinen mit EXAPT 2 (Determination of Optimal Cutting Values for Computer-Assisted Programming of N/C Lathes with EXAPT 2)," *Industrie-Anzeiger,* Vol. 90 (1968), No. 24, pp. 469–73.
14. Bernd Hirsch, "Ein System zur Ermittlung von Zerspanungsvorgabewerten, insbesondere bei rechnergestützte Programmierung numerisch gesteuerter Drehmaschinen (A System for the Determination of Cutting Values for the Computer-Assisted Programming of N/C Lathes)," doctoral thesis, University of Aachen, Aachen, Germany, 1968.
15. Bernd Hirsch and Hans Zölzer, *EXAPT 2 Material File,* Aachen, Germany, EXAPT-Verein, 1968.
16. Bernd Hirsch, "Automatic Programming for NC Lathes with EXAPT 2," *Machinery and Production Engineering,* August 6, 1969.
17. J. Witthoff, "Die Ermittlung der günstigsten Arbeitsbedingungen bei der spanabhebenden Formgebung (Determination of Optimal Machining Conditions for Metal Cutting)," *Werkstatt und Betrieb,* Vol. 85 (1952), No. 10, pp. 521–26.
18. J. Witthoff, "Ergänzende Betrachtungen zur Ermittlung der günstigsten Arbeitsbedingungen bei der spanabhebenden Formgebung (Additional Discussions of Determination of Optimal Machining Conditions for Metal Cutting)," *Werkstatt und Betrieb,* Vol. 90 (1957), No. 1, pp. 61–68.
19. W. Hans Engelskirchen, "Anpassung von Programmiersprachen der Fertigungstechnik an numerisch gesteuerte Werkzeugmaschinen (Adaptation of Production-Oriented Programming Languages to N/C Machine Tools)," doctoral thesis, University of Aachen, Aachen, Germany, 1968.
20. Michael Field, et al., "Computerized Determination and Analysis of Cost and Production Rates for Machining Operations," *ASME Paper No. 67-WA/PROD-18* (1967).
21. W. W. Gilbert and E. J. Weller, "Application of a Machinability Computer," *ASTE Annual Collected Papers,* Paper No. 24T26, 1956.
22. Inyong Ham, "Economics of Machining: Analyzing Optimum Machining Conditions by Computers," *ASTME Paper No. MR64-534* (1964).
23. *1130 Work Measurement Aids: User's Manual H20-0363-0,* White Plains, N.Y., International Business Machines Corporation, Technical Publications Department, 1967.

CHAPTER 4

COMPUTER ANALYSIS OF COST AND PRODUCTION RATE

Michael Field, Metcut Research Associates, Inc.
Roy L. Williams, Union Carbide Corporation
Alan F. Ackenhausen, Metcut Research Associates, Inc.

Technological developments in numerical control in the past few years have created a great deal of concern about the economic concepts of metal removal. Before numerical control, machine-tool time cost approximately $8 to $15 an hour; machine-tool time on N/C machines costs between $15 and $125 per hour, depending upon machine complexity. In the search for methods to reduce these high costs, increased efficiency of the actual chip-removal operation is generally most emphasized. While this emphasis is important, significant cost reductions can also be made in other phases of the machining cycle. A detailed cost analysis of total machine time will illuminate the areas where significant cost reductions can be made.

N/C demands that the process engineer or programmer supply proper information about cutting speeds, feeds, cutting tools, and fixture requirements for maximum N/C machine-tool efficiency. Data presented in this chapter show how each decision the engineer or programmer makes regarding the machining rates to be used can affect the total cost of the part.

The economic methods presented in this chapter enable the machining conditions for a part to be analyzed for economy by computer before they are committed to production. A cost breakdown such as those shown will indicate areas where improvement should be made to provide the maximum reduction in production cost.

COST AND PRODUCTION ANALYSIS METHODS

Three of the computerized programming systems discussed in the last chapter, the General Electric data system, the EXAPT system, and the Abex system, included special calculations that are intended to provide management with better control of the manufacturing process. These calculations, called high-efficiency or "optimum machining condition" calculations, are simply a means of controlling N/C machine tools to produce at the lowest part cost, the highest production rate, or some desired compromise of the two. In this chapter, computer analysis of cost and production rates is described in greater breadth and detail for all types of machining processes.

129

In the past, in most cases, the economics involved in the selection of machining parameters was based on engineering judgment of past experience with other parts rather than on careful cost analysis. The process engineer or parts programmer first selected the machining parameters to satisfy previously prescribed specifications for accuracy, surface finish, and overall surface integrity. When a part was machined with these quality specifications, a wide range of speeds, feeds, tool materials, and other machining conditions usually existed which could be used on a given machine.

A popular approach to this problem has been to apply an overall expression relating tool life to speed, feed, and other cutting parameters and to calculate optimum speed for those conditions (1, 2). One of the difficulties involved in this approach is that dependence on such relationships of tool life to speed and feed is unreliable and often erroneous (3). Optimum speed and feed obtained in this way are so general and approximate that their credence is questionable.

It has long been recognized that the cost and production rate can also be computed from actual shop data (4), but the equations needed for these computations, although simple, require long and laborious effort. The availability of the modern low-cost computer, however, has revived interest in this method of finding cost and production rates. The computer allows more conditions to be investigated for their effect in much less time. Third generation real-time and time-sharing computer systems can give the process engineer the solutions without the delays associated with batch processing. Now even rush jobs can be programmed economically. The computer has also revived interest in the determination of optimum machining conditions by employing empirical shop data (5, 6).

As described in Chapter 3, the first step in using a computer to find minimum cost and maximum production rate is to predetermine, by shop or laboratory testing, the relationship of tool life to cutting parameters for various machining operations on various alloys. These tool-life data, together with time-study data, can be inserted directly into the equations used in computing costs and production rates (7). The computer is not only programmed to print out the final cost and operation time, but also the individual factors which make up the final result. By analyzing the printout, one can rapidly separate significant from insignificant cost and production information.

MACHINABILITY DATA STORAGE AND RETRIEVAL

Significant machinability data relating tool life to the various machining parameters must be collected and organized in a usable form before machining costs, operation times, and production rates can be determined. Typical examples of data are the tables supplied by the Air Force Machinability Data Center (AFMDC) as shown in Figs. 4–1, 4–2, and 4–3 for turning, face milling, end-mill slotting, peripheral end milling, drilling, reaming, and tapping (8). Tool-life tests which simulate actual shop conditions should be made to determine the relationships between tool life and cutting speed for other machining conditions.

The recorded machinability data must next be key-punched on cards for computer processing. The use of punched cards allows greater flexibility in the handling of machining data. After a significant amount of data has been accumulated, a computerized file can be developed. In punch-card form the data can also be sorted into any desired sequence for listing. By storing the data in a random access file, inquiries for specific operations and materials or material groups can be made. The results of the file search are quickly printed out as shown in Figs. 4–4 and 4–5.

TURNING

MATERIAL	CONDITION & MICROSTRUCTURE	TOOL MATERIAL BHN	TRADE NAME	INDUSTRY GRADE	BR°	SR°	SCEA°	ECEA°	RELIEF°	NOSE RADIUS in.	CUTTING FLUID Code Appendix Page A-2	DEPTH OF CUT in.	FEED ipr	TOOL LIFE END POINT in.	TOOL LIFE - minutes vs SPEED-feet/minute R=Recommended Speed Appendix Page A-2
ALLOY STEELS - (cont.)															
8640	QUENCHED & TEMPERED TEMPERED MARTENSITE	400	-	T1 HSS	0	15	0	5	5	.005	11 1:20	.060	.009	.060	5/80 15/67 30/58 R
8640	ANNEALED 50%P-50%F	170	78	C-7	0	6	0	6	6	.040	00	.100	.010	.015	15/610 30/490 45/420 60/373 R
8640	SPHEROIDIZED SPHEROIDIZED CARBIDES + FERRITE	180	78	C-7	0	6	0	6	6	.040	00	.100	.010	.015	15/695 30/580 45/525 60/485 R
8640	ANNEALED 75%P-25%F	190	78	C-7	0	6	0	6	6	.040	00	.100	.010	.015	15/590 30/450 45/380 60/335 R
8640	ANNEALED WIDMANSTÄTTEN	250	78B	C-6	0	6	0	6	6	.040	00	.100	.010	.015	1/910 6/600 20/440 66/300
8640	ANNEALED WIDMANSTÄTTEN	250	78	C-7	0	6	0	6	6	.040	00	.100	.010	.015	15/640 30/520 45/440 60/375 R
8640	QUENCHED & TEMPERED TEMPERED MARTENSITE	300	78	C-7	0	6	0	6	6	.040	00	.100	.010	.015	5/660 15/400 20/315 R
8640	QUENCHED & TEMPERED TEMPERED MARTENSITE	400	78	C-7	0	6	0	6	6	.040	00	.100	.010	.015	5/480 15/365 20/318 R
52100	SPHEROIDIZED SPHEROIDIZED CARBIDES + FERRITE	190	-	T1 HSS	0	15	0	5	5	.005	11 1:20	.060	.009	.060	15/137 30/123 35/120 R
52100	SPHEROIDIZED SPHEROIDIZED CARBIDES + FERRITE	190	78B	C-6	0	6	0	6	6	.040	00	.100	.010	.015	15/430 30/340 45/300 R

Fig. 4-1.—Format for the collection of tool-life data for turning (8).

FACE MILLING

MATERIAL	CONDITION & MICROSTRUCTURE	BHN	TOOL MATL. TRADE NAME	TOOL MATL. INDUSTRY GRADE	UP OR DOWN MILLING	TOOL GEOMETRY AR	RR	CA	TR	INCL	ECEA	END REL	COR. REL	CUTTING FLUID Appendix p. A-2	DEPTH OF CUT in.	WIDTH OF CUT in.	FEED ipt	TOOL LIFE END POINT in.	TOOL LIFE/TOOTH inches work travel vs SPEED-feet/minute R=Recommended Speed Appendix Page A-2			
STAINLESS STEELS - MARTENSITIC																						
410	QUENCHED & TEMPERED, TEMPERED MARTENSITE	353	370	C-6	UP	0	-7	45	-5	5	5	8	8	00	.100	2.0	.010	.016	15 540	45 420	60 365	90 230 R
410	QUENCHED & TEMPERED, TEMPERED MARTENSITE	45 Rc	-	T15 HSS	UP	0	0	45	0	0	5	8	8	11 1:20	.060	2.0	.005	.060	15 100	30 79	50 68 R	

END MILL SLOTTING

MATERIAL	CONDITION & MICROSTRUCTURE	BHN	TOOL MATL. TRADE NAME	TOOL MATL. INDUSTRY GRADE	CUTTER TYPE	DIA. in.	NO. TEETH	FLUTE LENGTH in.	UP OR DOWN MILLING	TOOL GEOMETRY HELIX ANGLE°	RR°	CHAMFER	ECEA	CUTTING FLUID END REL	PERIPH. REL	Appendix p. A-2	DEPTH OF CUT in.	WIDTH OF CUT in.	FEED ipt	TOOL LIFE END POINT in.	TOOL LIFE/CUTTER inches work travel vs SPEED-feet/minute R=Recommended Speed Appendix Page A-2		
ALLOY STEELS																							
4340	SPHEROIDIZED CARBIDES, FERRITE	213	M2	HSS	SOLID HSS	.750	4	2	-	30	10	45 x .060	1	3	7	11 1.20	.250	.750	.002	.012	50 190	120 153	240 125

PERIPHERAL END MILLING

MATERIAL	CONDITION & MICROSTRUCTURE	BHN	TOOL MATL. TRADE NAME	TOOL MATL. INDUSTRY GRADE	CUTTER TYPE	DIA. in.	NO. TEETH	FLUTE LENGTH in.	UP OR DOWN MILLING	TOOL GEOMETRY HELIX ANGLE°	RR°	CHAMFER	ECEA	CUTTING FLUID END REL	PERIPH. REL	Appendix p. A-2	DEPTH OF CUT in.	WIDTH OF CUT in.	FEED ipt	TOOL LIFE END POINT in.	TOOL LIFE/CUTTER inches work travel vs SPEED-feet/minute R=Recommended Speed Appendix Page A-2		
ULTRA-HIGH STRENGTH STEELS																							
06AC	ANNEALED, 60 P-40 F	223	M2	HSS	SOLID HSS	.750	4	2	DOWN	30	10	45 x .060	1	3	7	11 1.20 MIST	.250	.750	.004	.012	75 290	130 240	260 190

Fig. 4-2.— Format for the collection of tool-life data for face milling, end-mill slotting, and peripheral end milling (8).

DRILLING

MATERIAL	CONDITION & MICROSTRUCTURE	DRILL MATL.			DRILL SIZE				DRILL GEOMETRY				CUTTING FLUID Appendix p. A-2	DEPTH OF HOLE in.	FEED ipr	DRILL LIFE END POINT in.	DRILL LIFE NO. OF HOLES vs SPEED-feet/minute R=Recommended Speed Appendix Page A-2			
		BHN	TRADE NAME	INDUS-TRY GRADE	TYPE DRILL	DIA. in.	LENGTH in.	FLUTE LENGTH in.	TYPE POINT	HELIX ANGLE°	POINT ANGLE°	LIP RE-LIEF°								
ALLOY STEELS -	QUENCHED & TEMPERED																25 98	50 84	75 76	100 70
4340	TEMPERED MARTENSITE	341	-	-	TWIST	.250	4.0	2.75	STANDARD	29	118	7	31	.5 THRU	.002	.015				
"	"	"			"	"	"	"	"	"	"	"	"	"	.005	"	25 80	50 65	75 56	100 50
"	"	"																		

REAMING

MATERIAL	CONDITION & MICROSTRUCTURE	BHN	REAMER DESCRIPTION				TOOL GEOMETRY				CUTTING FLUID Appendix p. A-2	STOCK ALLOW. ON DIA. in.	LENGTH OF HOLE in.	FEED ipr	TOOL LIFE END POINT in.	REAMER LIFE NO. OF HOLES vs SPEED-feet/minute R=Recommended Speed Appendix Page A-2	
			TOOL MATL. INDUSTRY TRADE NAME GRADE	DIA. in.	NO. OF FLUTES	STYLE	HELIX & HAND	CHAMFER	REL°								
ULTRA-HIGH STRENGTH STEELS -	ANNEALED & MARAGED	50R_c	M2 HSS	.272	6	CHUCKING	0	45 x .060	7	52	.022	.5 THRU	.005	.006	20 90	35 80	170 50
250 GRADE MARAGING STEEL	MARTENSITE						RH										

TAPPING

MATERIAL	CONDITION & MICROSTRUCTURE	BHN	TAP MATERIAL	TAP SIZE	NO. OF FLUTES	TAP STYLE	PERCENT OF THREAD	CUTTING FLUID Appendix p. A-2	DEPTH OF HOLE in.	TAP LIFE END POINT	TAP LIFE NO. OF HOLES vs SPEED-feet/minute R=Recommended Speed Appendix Page A-2		
HIGH TEMPERATURE ALLOYS - IRON BASE													
WROUGHT A-286	SOLUTION TREATED & AGED AUSTENITIC	320	M10 HSS	5 16-18NC	2	PLUG SPIRAL POINT	75	53	.5 THRU	TAP BREAKAGE	12 50	31 40	132 30 R

Fig. 4-3.—Format for the collection of tool-life data for drilling, reaming, and tapping (8).

LISTING OF N/C MACHINING DATA FROM COMPUTERIZED FILE

TURNING

MATERIAL	CONDITION &—— BHN *MICRO* *STRUCTURE*	TOOL MAT *TRAD*IND* *NAME*GRD	TOOL GEOMETRY BR A SR SGEA ECEA REL N.R.		DEPTH CUT. OF FLD. CUT *FEED (IN) (IPR)	WEAR LAND (IN)	TOOL LIFE (MIN) SPEED (FT/MIN) R=RECOMMENDED SPEED			SOURCE VS VOL REF		

FREE MACHINING PLAIN CARBON STEELS

| B1112 | ANNEALED 135 10P-90F+S | T1 HSS | 0 15 0 5 5 | .005 11 1/20 | .0600 | .0090 .060 | 8 15 30 300 270 240R | 0 0 | 0 2 36 |
| B1112 | ANNEALED 135 788 10P-90F+S | C-6 | 0 6 0 6 6 | .040 0 | .1000 | .0100 .015 | 15 30 45 999 920 860 | 60 810R | 0 2 36 0 |

PLAIN CARBON STEELS

1020	ANNEALED 116 10P-90F+S	T1 HSS	0 15 0 5 5	.005 11 1/20	.0600	.0090 .060	288 260 237	0 0	0 2 40
1020	ANNEALED 115 788 10P-90F+S	C-6	0 6 0 6 6	.040 0	.1000	.0100 .015	999 960 800	0 0	0 2 40
1020	ANNEALED 115 78 10P-90F+S	C-7	0 6 0 6 6	.040 0	.1000	.0100 .015	15 30 45 999 880 740	60 630R	0 2 40

FREE MACHINING ALLOY STEELS

3140RE-S	ANNEALED 190 75P-25F+S	T1 HSS	0 15 0 5 5	.005 11 1/20	.0600	.0090 .060	15 30 45 140 130 126	60 123R	0 2 54 0
3140RE-S	Q & T 296 TMP-MRT+S	T1 HSS	0 15 0 5 5	.005 11 1/20	.0600	.0090 .060	15 30 45 109 106 105R	0 0	0 2 56
3140RE-S	ANNEALED 190 788 75P-25F+S	C-6	0 6 0 6 6	.040 0	.1000	.0100 .015	15 30 45 605 485 415	60 360R	0 2 54 0

Fig. 4-4.—Sample computerized file listing of N/C machining data for turning. Note the similarity of the data format to that in Fig. 4-1.

One method of sorting data for a given operation is to arrange the data by families of work materials and in order of increasing work material hardness. Fig. 4-4 shows a partial computer listing for the turning of various steels. Note that it is similar to Fig. 4-1 in arrangement. Fig. 4-5 shows a partial computer listing of data for the drilling of alloy steels, with a data format similar to that of Fig. 4-3. These data may be changed or added to at any time. Additional or revised data cards can be merged with the master deck for printing of a master manuscript, which can then be photoprinted for distribution.

Within most manufacturing plants, significant machinability data can be gathered from actual machining operations, provided that a simplified form which allows the operator to record a minimum amount of data is used. The operator can then enter the tool number to describe the tool geometry, the part number to describe the work material, and other numbers for additional data. This approach is particularly valuable if a company already has a computer file system to keep track of work in progress, operator's time, etc. The numbers are then decoded from those files that contain the necessary information. The machining data can also be combined with whatever other data is recorded from remote control terminals. If a data collection system does not already exist, however, the machining data may be written on forms or data cards for periodical collection and then be key-punched, listed for analysis, and eventually stored in a file where they will then be available for future reference and analysis.

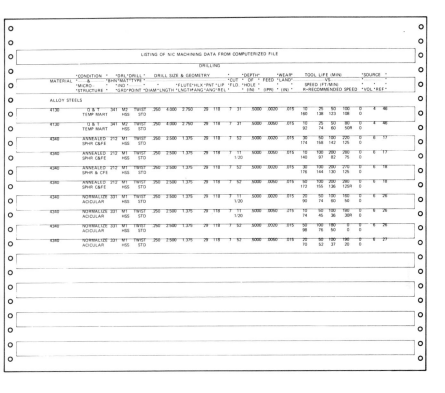

Fig. 4–5.—Sample computerized file listing of N/C machining data for drilling. Note the similarity of the data format to that in Fig. 4–3.

Data Requirements

Machining data requirements may be divided into the following three groups:

1) Machinability parameters
2) Time-study and cost data
3) Items relative to part geometry and tool motion on the machine.

The symbols used to represent the various elements of the data requirements discussed in this section, with their definitions and the ways they apply to various operations, are shown in Table IV–1.

Machinability Parameters. Machinability data for a specific material, and for its particular condition and hardness, should include the following:

1) Relationship of tool life to cutting speed
2) Feed rate
3) Number of cutter teeth (milling) or number of threads (tapping)
4) Tool material
5) Tool geometry
6) Cutting fluid
7) Depth of cut
8) Width of cut
9) Tool-life end point.

Table IV-1. Symbols for Cost and Production Rate Equations for N/C Machining.

Symbol	Definition	Turn	Mill	Drill and Ream	Tap	Center Drill
C	cost for machining one workpiece, $/workpiece	✓	✓	✓	✓	✓
C_c	cost of each insert or inserted blade, $/blade	✓	✓	No	No	No
C_p	purchase cost of tool or cutter, $/cutter	✓	✓	✓	✓	✓
C_w	cost of grinding wheel for resharpening tool or cutter, $/cutter	✓	✓	No	No	No
d	depth of cut, in.	✓	✓	No	No	No
D	dia. of work in turning, of tool in milling, drilling, reaming, tapping, in.	✓	✓	✓	✓	✓
e	extra travel at feedrate (f_r or f_t) including approach, overtravel, and all positioning moves, in.	✓	✓	✓	✓	✓
f_r	feed per revolution, ipr	✓	No	✓	No	✓
f_t	feed per tooth, in./tooth	No	✓	No	No	No
G	labor plus overhead in tool reconditioning department, $/min	✓	✓	✓	✓	✓
k_1	no. times lathe tool, milling cutter, drill, reamer, or tap is resharpened before being discarded	✓	✓	✓	✓	✓
k_2	no. times lathe tool or milling cutter is resharpened before inserts or blades are rebrazed or reset	✓	✓	No	No	No
k_3	no. times blades (or inserts) are resharpened (or indexed) before blades (or inserts) are discarded	✓	✓	No	No	No
L	length of workpiece in turning and milling or sum of length of all holes of same diameter in drilling, reaming, or tapping, in.	✓	✓	✓	✓	✓
m	no. threads per inch	No	No	No	✓	No
M	labor plus overhead cost on lathe, milling machine, or drilling machine; $/min	✓	✓	✓	✓	✓
n	tool-life exponent in Taylor's equation (Eq. 3–2)	✓	✓	✓	✓	No
N_L	no. workpieces in lot	✓	✓	✓	✓	✓
P	production rate per 60 min. workpieces/hr	✓	✓	✓	✓	✓
r	rapid traverse rate, in./min	✓	✓	✓	✓	✓
R	total rapid traverse distance for a tool or cutter on one part, in.	✓	✓	✓	✓	✓
S	reference cutting speed for a tool life of $T = 1$ min, sfpm	✓	No	No	No	No
S_t	reference cutting speed for a tool life of $T_t = 1$ in., sfpm	No	✓	✓	✓	No
t_b	time to rebraze lathe tool or cutter teeth or reset blades, min	✓	✓	No	No	No
t_d	time to replace dull cutter in tool changer storage unit, min	✓	✓	✓	✓	✓

Table IV-1. Symbols for Cost and Production Rate Equations for N/C Machining. (*Continued*)

Symbol	Definition	Applies to Operation				
		Turn	Mill	Drill and Ream	Tap	Center Drill
t_i	time to index from one type cutter to another between operations (automatic or manual), min	✓	✓	✓	✓	✓
t_L	time to load and unload workpiece, min	✓	✓	✓	✓	✓
t_m	time (average) to complete one operation, min	✓	✓	✓	✓	✓
t_o	time to set up machine tool for operation, min	✓	✓	✓	✓	✓
t_p	time to preset tools away from machine (in toolroom), min	✓	✓	✓	✓	✓
t_s	time to resharpen lathe tool, milling cutter, drill, reamer, or tap, min/tool	✓	✓	✓	✓	✓
T	tool life measured in minutes to dull a lathe tool, min	✓	No	No	No	No
T_h	no. holes per resharpening	No	No	No	No	✓
T_t	tool life measured in inches travel of work or tool to dull a drill, reamer, tap, or one milling cutter tooth, in.	No	✓	✓	✓	No
u_c	no. holes center drilled of chamfered in workpiece	No	No	No	No	✓
v	cutting speed, sfpm	✓	✓	✓	✓	✓
w	width of cut, in.	No	✓	No	No	No
z	no. teeth in milling cutter or no. flutes in tap	No	✓	No	✓	No

These data are shown in Figs. 4–1, 4–2, and 4–3. Of these parameters, the feed rate, the tool life vs. cutting speed, and the number of cutter teeth in milling or the number of threads per inch in tapping are used directly in calculating costs and operation times. The tool material, tool geometry, cutting fluid, tool-life end point, and depth and width of cut, where applicable, are indirectly involved, since they all have an effect on tool life. For this reason the machinability parameters must be considered as a set of conditions, and a numerical identification is necessary for each data set.

Time-Study and Cost Data. The time-study and cost data comprise the costs associated with the machine tool and the costs associated with tool or cutter reconditioning. The machine-tool cost elements consist of the following items:

1) Labor and overhead rate for the machine and operator
2) Tool index time
3) Time required to replace dull tools
4) Machine setup time
5) Time required to load and unload the workpiece.

The tool reconditioning cost elements include the following:

1) Labor and overhead rate in the toolroom
2) Purchase cost of the tool

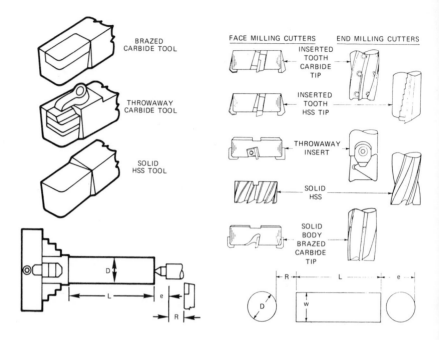

Fig. 4–6.—Lathe tools and their setup corresponding to Eq. 4–1, Table IV-2.

Fig. 4–7.—Milling cutters and their setup corresponding to Eq. 4–2, Table IV-2.

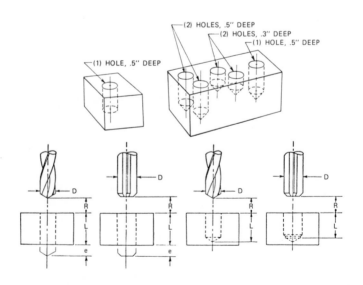

Fig. 4–8.—Drills, reamers, and their setup corresponding to Eq. 4–3, Table IV-2.

3) Cost of the insert
4) Grinding wheel cost for resharpening the tool
5) Number of times the tool is resharpened before being discarded
6) Number of times the tool is resharpened before the inserts or blades are rebrazed or reset
7) Number of times the blades are resharpened or inserts are indexed before the blades or inserts are discarded
8) Time required for presetting the tools
9) Time required to resharpen the tool.

Part Geometry and Tool Motion Data. The following data are related to part geometry and tool moves:

1) Diameter of the tool or workpiece
2) Depth of cut
3) Length of cut
4) Width of cut
5) Extra travel of the tool in feed but not cutting
6) Distance traveled in rapid traverse
7) Rapid traverse rate.

The dimensions listed, with the respective types of tools, are described in Fig. 4–6 for turning, Fig. 4–7 for milling, Fig. 4–8 for drilling and reaming, and Fig. 4–9 for tapping. The symbols assigned to the dimensions are those defined in Table IV–1 and used in the cost and production equations later in this chapter. Although these symbols differ in some cases from those used in Chapter 3, this difference is due only to differences in terminology of the computer programming models used in each chapter.

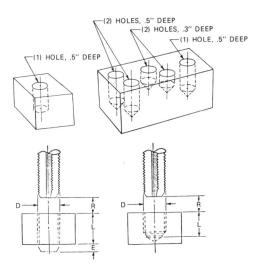

Fig. 4–9.—Taps and their setup corresponding
to Eq. 4–4, Table IV–2.

Table IV-2. Generalized Cost

Turning

$$C = M \left[\frac{D(L+e)}{3.82\, f_r v} + \frac{R}{r} \qquad\qquad + \quad t_i \quad + \quad \frac{D L\, t_d}{3.82\, f_r v\, T} \right]$$

Milling

$$C = M \left[\frac{D(L+e)}{3.82\, Z f_t v} + \frac{R}{r} \qquad\qquad + \quad t_i \quad + \quad \frac{L\, t_d}{Z T_t} \right]$$

Drilling or Reaming

$$C = M \left[\frac{D(L+e)}{3.82\, f_r v} + \frac{R}{r} \qquad\qquad + \quad t_i \quad + \quad \frac{L\, t_d}{T_t} \right]$$

Tapping

$$C = M \left[\frac{m\, D(L+e)}{1.91\, v} \quad \frac{R}{r} \qquad\qquad + \quad t_i \quad + \quad \frac{L\, t_d}{T_t} \right]$$

Center Drilling or Chamfering

$$C = M \left[\frac{D(L+e)}{3.82\, f_r v} + \frac{R}{r} \qquad\qquad + \quad t_i \quad + \quad \frac{u_c l_d}{T_h} \right]$$

Handling and Setup

$$C = M \left[\qquad\qquad\qquad\qquad\qquad\qquad t_L \quad + \quad \frac{t_0}{N_L} \qquad\qquad\qquad\qquad\qquad \right]$$

$\dfrac{\$}{Min}$	Feeding Time	Rapid Traverse Time	Load & Unload Time	Setup Time	Cutter Index Time	Dull Tool Replacement Time

Data Input

The data input is keypunched on cards. The first card contains part information —the part number and name, material, hardness, and lot size. Machining parameters, time-study and cost data, and part geometry data are key-punched on a set of two cards and are identified by a data-set number. All values required for each calculation, except lot size, are included in the two-card set. By changing the lot-size quantity in the part information card and by reusing the data-set cards, the effect of lot size on costs can be analyzed. Data sets for various operations and types of tools may be entered in any desired sequence, but are usually inserted in the order of part processing. Within one operation, the various cutting speed vs. tool life data may be in a sequence corresponding to decreasing cutting speed.

COMPUTATION OF COST AND OPERATION TIMES

Machining cost per piece consists of the total of the average costs per piece for each operation plus handling cost (loading and unloading costs) plus prorated setup cost (setup costs divided by lot size). Production rate is simply 60 min divided by the summation of all the operation times plus handling time plus prorated setup time.

Equations for N/C Machining.

$$+ \frac{D L}{3.82\, f_r v\, T}\left[\frac{C_p}{(k_1 + 1)} + G t_s + \frac{G t_b}{k_2} + \frac{C_c}{k_3} + C_w + G t_p\right] \quad (4\text{-}1)$$

$$+ \frac{L}{\mathcal{Z} T_t}\left[\frac{C_p}{(k_1 + 1)} + G t_s + \frac{G t_b}{k_2} + \frac{\mathcal{Z} C_c}{k_3} + C_w + G t_p\right] \quad (4\text{-}2)$$

$$+ \frac{L}{T_t}\left[\frac{C_p}{(k_1 + 1)} + G t_s \qquad\qquad\qquad\qquad\qquad + G t_p\right] \quad (4\text{-}3)$$

$$+ \frac{L}{T_t}\left[\frac{C_p}{(k_1 + 1)} + G t_s \qquad\qquad\qquad\qquad\qquad + G t_p\right] \quad (4\text{-}4)$$

$$+ \frac{u_c}{T_h}\left[\frac{C_p}{(k_1 + 1)} + G t_s \qquad\qquad\qquad\qquad\qquad + G t_p\right] \quad (4\text{-}5)$$

$$(4\text{-}6)$$

↑	↑	↑	↑	↑	↑
Tool Depreciation Cost	Tool Resharpening Cost	Rebrazing or Blade Reset Cost	Insert or Blade Cost	Grinding Wheel Cost	Tool Presetting Cost

To find operation machining costs and operation times, the computer solves several generalized machining cost and operation time equations as shown in Tables IV–2 and IV–3. Eqs. 4–1 through 4–5 in Table IV–2 give the average cost for each operation—turning, milling, drilling or reaming, tapping, and center drilling or chamfering, respectively. Each of the operation costs consists of the cost factors associated with the operation of the machine tool and with tool or cutter reconditioning costs, as shown by the brackets in the equations. The handling-time and prorated-setup-time cost factors have been separated from the machine-tool cost factors as shown in Eq. 4–6.

A corresponding set of generalized equations for operation time, Table IV–3, contains Eq. 4–7 for turning, Eq. 4–8 for milling (face milling, peripheral end milling, or end-mill slotting), Eq. 4–9 for drilling or reaming, Eq. 4–10 for tapping, and Eq. 4–11 for center drilling or chamfering. The equation for production rate in pieces per hour is shown as Eq. 4–12.

The generalized N/C machining cost and operation time equations are arranged in the tables so that the corresponding cost factors of the various operations are aligned vertically. For example, the first term in the first bracket in each equation indicates feed time, the second term indicates rapid traverse time, and so on for each operation. The symbols used for all the equations are those defined in Table IV–1. These equations are based on the assumption that one tool is used for one and only one operation.

**Table IV-3. Generalized Equations for Operation Time per Piece
and Production Rate for N/C Machine Tools.**

Turning

$$t_m = \frac{D(L+e)}{3.82 \, f_r \, v} + \frac{R}{r} + t_i + \frac{D \, L \, t_d}{3.82 \, f_r \, v \, T} \tag{4-7}$$

Milling

$$t_m = \frac{D(L+e)}{3.82 \, Z \, f_t v} + \frac{R}{r} + t_i + \frac{L \, t_d}{Z \, T_t} \tag{4-8}$$

Drilling or Reaming

$$t_m = \frac{D(L+e)}{3.82 \, f_r v} + \frac{R}{r} + t_i + \frac{L \, t_d}{T_t} \tag{4-9}$$

Tapping

$$t_m = \frac{m D(L+e)}{1.91 \, v} + \frac{R}{r} + t_i + \frac{L \, t_d}{T_t} \tag{4-10}$$

Center Drilling or Chamfering

$$t_m = \frac{D(L+e)}{3.82 \, f_r v} + \frac{R}{r} + t_i + \frac{u_c \, t_d}{T_h} \tag{4-11}$$

Production Rate

$$P = \frac{60}{\Sigma \, t_m + \left(t_L + \dfrac{t_o}{N_L} \right)} \tag{4-12}$$

This assumption avoids complicating the relationship between length of cut and tool life. All machining sequences using the same tool should be considered as a single operation.

The generalized cost equations could be broken down and adapted to the specific type of tool used for each operation. This breakdown is unnecessary, however, since the nonapplicable terms of the tool or cutter reconditioning cost factors are omitted by the computer program. The operation and type of tool involved are determined by the computer from a two-digit code, and only the variables that apply to the machining situation need to be given in the input data. As shown in Table IV–2, blade reset and blade (or insert) cost factors do not apply when solid HSS tools or cutters are used for turning or milling. In the same way, when throw-away carbide insert tools or cutters are used in turning or milling, tool resharpening, blade reset, and grinding wheel cost factors are not applicable. However, when brazed carbide tools are used in turning, or when solid-body brazed carbide tip cutters or inserted-tooth cutters (carbide tip or HSS blade) are used for milling, all cost factors are applicable. For drilling or reaming, tapping, or center drilling or chamfering, the blade reset and the insert cost factors do not apply, and the grinding wheel cost factor is negligible.

Computer Output

The results of the computer's calculations are printed out as shown in Fig. 4–10. The main heading is followed by the part number and name, lot size, work material,

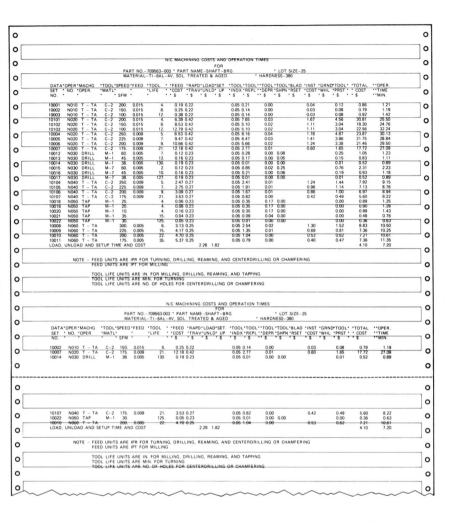

Fig. 4–10.—Computer printout of N/C machining costs and operation times for Example 1, a shaft to be machined on an N/C lathe.

and hardness. The subheadings are then printed followed by the data in the order of input. Each subheading includes the data-set number, the operation number, a mnemonic code for the machining operation (including tool type, where applicable), tool material, cutting speed, feed rate, tool life, twelve cost factors, and the computed total cost and operation time for each data set. The data-set number refers to the other machining parameters not specifically listed, such as tool geometry, cutting fluid, and depth and width of cut. The load, unload, and prorated setup time and cost are listed following the last operation line. A note indicating feed and tool-life units is printed out, and finally the analyzed results for each operation, to be discussed in the next section, are printed out to complete the listing.

COST AND PRODUCTION ANALYSIS

As mentioned earlier, machinability data may be entered in the order of the operation number corresponding to the order of processing, and various sets of parameters can be investigated for each operation. The computer printout of the computed results (see Fig. 4–10) permits analysis of each individual cost factor as well as total operation cost and operation time for each data set. From the various conditions calculated for each operation, the computer selects the set of machining parameters that gives the lowest cost or the shortest operation time (for the maximum production rate).

Some compromise between these two items might be made. If a minimum cost or operation time is not evident because it lies beyond the range of the data investigated and entered into the computer, then calculations can be made to determine the cutting speed and tool life for either a minimum-cost or maximum-production projection. These calculations are discussed later in this chapter. Any projections of this type should be verified by tests before they are used in production.

After the machining conditions for each operation have been selected, the total costs for the operations are added to handling and prorated setup cost to arrive at the total cost per piece. If the minimum operation costs are selected, the minimum cost per piece is the result. Addition of the selected operation times to the handling and prorated setup time gives the total time in minutes for each part. When divided into 60 min as shown in Eq. 4–12, total per-part time yields the production rate in pieces per hour. Selection of the minimum operation times yields the minimum total time and therefore the maximum production rate. Two examples of cost and production analysis follow:

Example 1: An N/C Lathe

The part shown in Fig. 4–11 is to be machined from 4.375-in.-diameter bar stock. The material is Ti-6Al-4V alloy which is solution treated and aged to a hardness of 380 BHN. The bar stock is chucked and the part is rough and finish turned and faced, drilled, and tapped complete on one end. Another setup is required to finish the part, but for brevity it is not considered here.

Operation 10, the first operation, consists of rough facing the end of the stock. Rough turning of the stock is done in operation 20 with one 8.5-in.-long cut and five 5-in.-long cuts, each .100 in. deep. A .25-in. stub drill is fed in for operation 30. Semifinish profiling is done in operation 40. The hole in the end of the part is tapped with a floating tap driver in operation 50, and the part is then finish turned and faced com-

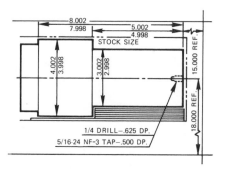

Fig. 4–11.—Part described in Example 1, a shaft to be machined on an N/C lathe.

plete in operation 60. Each operation requires an index, since six tools are used on a hexagonal turret block. The depth and length of cut, index time, extra travel in feed, and total distance traveled in rapid traverse are given in Fig. 4–12 for each operation. Throw-away carbide insert tools are used for the turning and facing operations.

Fig. 4–12 also gives the cost and time-study data for each operation. Fig. 4–13 gives the tool-life vs. cutting-speed, feed-rate, tool-material, and tool-geometry data for each of the operations. The circled numbers identify the tool-life vs. cutting-speed data sets and are used as the fourth and fifth digits of the five-digit data-set numbers to identify tool-life data in the computer printout (Fig. 4–10). When tool life vs. speed data are used twice, a 1 is printed as the third digit of the data-set number for the second usage.

Three sets of data were analyzed for the rough-facing operation, N010 (Fig. 4–10). As the cutting speed decreased, the tool life increased, while the total operation cost first decreased and then increased. For data set 10002, the cost is $.79 and the operation time is 1.19 min, both of which are minimum values for the data investigated. Using the same tool-life data with a feed of .015 ipr for the rough-turning operation, N020, shows similar results, with the 150-sfpm cutting speed yielding the lowest cost of $18.30. However, with a .009-ipr feed and a cutting speed of 175 sfpm, data set 10007 yields a still lower cost of $17.72 for the operation but increases the operation time from 24.76 to 27.09 min. A slower speed may be worth investigating, for the minimum cost has not yet been reached.

The drilling operation, N030, illustrates the use of two different drill materials, M1 and M7, but the minimum cost of both is $.52, and the operation time is also held at a minimum for both. Note the wide variation in tool life and cost represented by a decrease in cutting speed of only 12 sfpm. The conditions for data set 10014 would be selected because the M1 drills are more readily available than the M7 drills.

The semifinish profile turning operation, N040, shows decreasing cost and time values as the cutting speed decreases to 175 sfpm. A further decrease in cutting speed should be investigated, although the feed cost factor is increasing and may overtake the reductions in the other cost factors. The tapping operation, N050, compares two-flute taps (data sets 10018 to 10020) to three-flute taps (data sets 10021 and 10022). The three-flute taps give a lower cost and operation time due to the longer tool life at higher speeds. Data set 10022 gives the lowest cost of $.36 and operation time of .63 min.

The finish profile-turning operation, N060, gives ideal results. The operation cost passes through a minimum of $7.21 for data set 10010, while the minimum operation time appears to be near the conditions given by data set 10009. The total handling and prorated setup costs and times are given as $4.10 and 7.20 min, respectively.

We can now total the minimum operation costs and the load, unload, and prorated setup costs of the chosen data sets.

Operation	Cost
10002	$ .79
10007	17.72
10014	.52
10107	5.60
10022	.36
10010	7.21
Load, unload, and setup	4.10
Total part cost	$36.30

Oper. No.	Operation	Part No. 543109-001					
		Part Name Shaft					
N 010	**Rough face end**						
N 020	**Rough turn**	**Material** Ti-6Al-4V					
N 030	**Drill hole—.250-in. dia.**						
N 040	**Semifinish profile**	**Condition** Solution-treated, Aged					
N 050	**Tap hole—.250-in. tap**						
N 060	**Finish turn and face**	**Hardness** 380 BHN					
		Operation No.					
		N 010	N 020	N 030	N 040	N 050	N 060
C_c	cost of each carbide tip or insert, $	4.99	4.99	——	5.85	——	5.85
C_p	purchase cost of tool, $	29.33	29.33	†	46.06	1.40	46.06
C_w	cost of grinding wheel for resharpening tool, $	——	——	——	——	——	——
d	depth of cut, in.	.100	.100	——	.060	——	.027
D	diameter of workpiece in turning, of tool (avg.) in drilling and tapping	1.625	3.675	.250	3.555	.312	3.000
e	extra travel at feedrate including overtravel and all positioning moves, in.	.250	1.500	.250	.250	.250	.250
f_r	feed per revolution, ipr	*	*	*	*	——	*
G	labor plus overhead in tool reconditioning dept., $/min	.17	.17	.17	.17	.17	.17
k_1	no. times tool is resharpened before discarding, or no. times insert is indexed before discarding tool holder	2,000	2,000	12	2,000	——	2,000
k_2	no. times tool is resharpened before rebrazing or resetting	——	——	——	——	——	——
k_3	no. times insert is resharpened (or indexed) before insert is discarded	8	8	——	4	——	4
L	length of cuts on workpiece at the same speed using the same tool, in.	2.125	33.5	.75	10.25	.500	10.25
m	no. of threads per in.	——	——	——	——	24	——
M	labor and overhead cost on lathe, $/min	.57	.57	.57	.57	.57	.57
N_L	no. of pieces in lot	25	25	25	25	25	25
r	rapid traverse rate, in./min	100	100	100	100	100	100
R	total rapid traverse distance for a tool on one part, in.	39.73	74.77	41.97	47.39	41.34	43.89
t_b	time to rebraze tool, min	——	——	——	——	——	——
t_d	time to replace dull tool in turret block, min	5	5	5	5	5	5
t_i	time to index turret from one tool to another between operations, min	.1	.1	.1	.1	.1	.1
t_L	time to load and unload workpiece, min	4	4	4	4	4	4
t_o	time to set up lathe for operation, min	80	80	80	80	80	80
t_p	time to preset tools away from machine (in tool room), min	10	10	15	10	10	10
t_s	time to resharpen tool, min	——	——	5	——	——	——
T	tool life, total time to dull tool, min	*	*	——	*	——	*
T_t	tool life in in. travel to dull a drill or tap			*		*	
v	cutting speed, sfpm	*	*	*	*	*	*

*These values taken from tool life data, Fig. 4–13
†Cost of M1 drill is $.60; cost of M7 HSS drill is $.90

Fig. 4–12.—Time-study and cost data for Example 1.

TURNING

MATERIAL	CONDITION & MICROSTRUCTURE	BHN	TOOL MATERIAL TRADE NAME	TOOL MATERIAL INDUSTRY GRADE	BR°	SR°	SCEA°	ECEA°	RELIEF°	NOSE RADIUS in.	CUTTING FLUID Code*	DEPTH OF CUT in.	FEED ipr	TOOL LIFE END POINT in.	TOOL LIFE – minutes vs SPEED-feet/minute
TITANIUM ALLOYS Ti-6Al-4V	SOLUTION TREATED & AGED ALPHA-BETA	388	883	C-2	-5	-5	15	15	5	.030	00	.062	.005	.015	⑧ 6 / 300, ⑨ 9 / 225, ⑩ 15 / 200(?), ⑪ 33 / 175
"	"		"	"	"	"	"	"	"	"	"	"	.009	"	④ 5 / 250, ⑤ 7 / 225(?), ⑥ 9 / 200, ⑦ 21 / 175
"	"		"	"	"	"	"	"	"	"	"	"	.015	"	① 4 / 200, ② 8 / 150, ③ 22 / 100

DRILLING

MATERIAL	CONDITION & MICROSTRUCTURE	BHN	DRILL MATL. TRADE NAME	DRILL MATL. INDUSTRY GRADE	TYPE DRILL	DIA. in.	LENGTH in.	FLUTE LENGTH in.	TYPE POINT	HELIX ANGLE°	POINT ANGLE°	LIP RELIEF°	CUTTING FLUID Code*	DEPTH OF HOLE in.	FEED ipr	DRILL LIFE END POINT in.	DRILL LIFE NO. OF HOLES vs SPEED-feet/minute
TITANIUM ALLOYS Ti-6Al-4V	SOLUTION TREATED & AGED ALPHA-BETA	375	-	M1 HSS	TWIST	.250	2.5	1.375	CRANK-SHAFT	29	118	7	53 FLOOD	.500 THRU	.005	.015	㉒ 15 / 60, ㉓ 25 / 45, ㉔ 250 / 38
"	ALPHA-BETA		M7 HSS		TWIST OIL HOLE	"	"	"	"	15	"	18	"	"	"	"	5 / 60, 20 / 45, 255 / 38

TAPPING

MATERIAL	CONDITION & MICROSTRUCTURE	BHN	TAP MATERIAL	TAP SIZE	NO. OF FLUTES	TAP STYLE	PERCENT OF THREAD	CUTTING FLUID Code*	DEPTH OF HOLE in.	TAP LIFE END POINT in.	TAP LIFE NO. OF HOLES vs SPEED-feet/minute
TITANIUM ALLOYS Ti-6Al-4V	SOLUTION TREATED & AGED ALPHA	375	M1 HSS	5/16-24NF	2	PLUG SPIRAL POINT	75	53	.5 THRU	UNDER-SIZE THREAD	⑱ 8 / 25, ⑲ 20 / 20, ⑳ 8 / 10
"	"			"	3						㉒ 30 / 250, 35 / 30(?)

*Cutting Fluid Code   00 Dry   53 Highly Chlorinated Mineral Oil

Fig. 4-13.—Tool-life data for turning, drilling, and tapping a part on an N/C lathe.

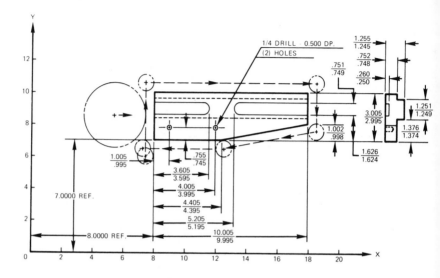

Fig. 4–14.—Part described in Example 2 to be machined on an N/C machining center.

The corresponding operation times total 55.83 min. Dividing this total average floor-to-floor time into 60 min gives us a production rate of 1.076 pieces/hr.

Example 2: An N/C Machining Center

The part to be finish machined, shown in Fig. 4–14, is a forging which has had its bottom surfaces finish machined previously. The material is 4340 alloy steel in the normalized condition with a hardness of 331 BHN. The part is clamped in a vise across its 1.251/1.249-in. dimension. An N/C machining center with an automatic tool changer is used.

Operation 10 consists of face milling the top using a 4-in.-diameter cutter with throw-away carbide inserts. An HSS peripheral end mill is used to mill the part to the correct width and length in operation 20. The two slots are milled in operation 30 with an HSS end mill. Two holes are then drilled in operation 40 to complete the part. The cutting dimensions (depth, width, and length of cut), extra travel in feed, total distance traveled in rapid traverse, and change time are given in Fig. 4–15. Fig. 4–15 also lists the cost and time-study data for each operation.

Fig. 4–16 gives the tool-life vs. speed, feed, tool-material, and tool-geometry data for the various machining operations. The circled numbers, as in the previous example, are used to identify the tool-life-vs.-cutting-speed data sets. These numbers are used as the fourth and fifth digits of the five-digit data set numbers shown in Fig. 4–17. The sets of data giving the drill life in terms of the number of holes vs. speed are numbered 12 through 19. The equations require the drill life, T_t, to be expressed in inches. Therefore, it is necessary to multiply the drill life (number of holes drilled) by the hole length to get the drill life measured in inches of travel required to dull the drill. For example, using data set 12 from Fig. 4–16, the drill life is 20 holes, and the length of each hole in Fig. 4–14 is .5 in. Therefore, the drill life is 10 in. ($T_t = 20 \times .5$ in.).

The computer printout of the calculated results is shown in Fig. 4–17. Four cutting speeds are considered for the face-milling operation, N010. The first data set, number

Oper. No.	Operation	Part No. 456987-002			
		Part Name Bracket—Slider			
N 010	Face mill top surface	**Material** AISI 4340 Steel Forg.			
N 020	Peripheral end mill sides and ends	**Condition** Normalized			
N 030	End mill two slots in top surface				
N 040	Drill two holes—.250-in. dia.	**Hardness** 331 BHN			
		Operation No.			
		N 010	N 020	N 030	N 040
C_c	cost of each insert or inserted blade, $/blade	2.45	——	——	——
C_p	purchase cost of tool or cutter, $/cutter	212.00	10.00	6.40	0.60
C_w	cost of grinding wheel for resharpening tool or cutter, $/cutter	——	0.10	.10	——
d	depth of cut, in.	0.100	0.250	0.250	——
D	dia. of tool in milling and drilling, in.	4.000	1.000	0.750	0.250
e	extra travel at feedrate (f_r or f_t) including approach, overtravel, and all positioning moves, in.	5.00	5.50	2.25	0.50
f_r	feed per revolution, ipr	——	——	——	*
f_t	feed per tooth, in./tooth	*	*	*	——
G	labor plus overhead in tool reconditioning department, $/min	0.07	0.17	0.17	0.17
k_1	no. times milling cutter or drill is reused before being discarded	9000	7	7	12
k_2	no. times milling cutter is resharpened before inserts (or blades) are rebraced (or reset)	——	——	——	——
k_3	no. times blades (or inserts) are resharpened (or indexed) before blade (or inserts) are discarded	8	——	——	——
L	length of workpiece in milling or sum of length of all holes of same diameter in drilling, in.	10.00	26.25	8.40	1.00
M	labor plus overhead cost of machining center, $/min	0.83	0.83	0.83	0.83
N_L	no. of workpieces in lot	20	20	20	20
r	rapid traverse rate, in./min	150	150	150	150
R	total rapid traverse distance for a cutter on one part, in.	43.00	27.25	46.35	41.00
t_b	time to rebraze cutter teeth or reset blades, min	——	——	——	——
t_d	time to replace dull cutter in tool changer storage unit, min	3.0	3.0	3.0	3.0
t_i	time to index from one type cutter to another between operations (automatic or manual), min	0.1	0.1	0.1	0.1
t_L	time to load and unload workpiece, min	2.0	2.0	2.0	2.0
t_o	time to set up machine tool for operation, min	60	60	60	60
t_p	time to preset tools away from machine (in toolroom), min	20	15	15	12
t_s	time to resharpen lathe tool, milling cutter, drill, reamer, or tap, min/tool	——	15	15	5
T_t	tool life measured in inches travel of work or tool to dull a drill, reamer, tap, or one milling cutter tooth, in.	*	*	*	*
v	cutting speed, sfpm	*	*	*	*
w	width of cut, in.	3.50	0.75	0.75	——
Z	no. teeth in milling cutter	6	4	4	——

*These values taken from tool life data, Fig. 4–16.

Fig. 4–15.—Time-study and cost data for Example 2.

FACE MILLING *

*Single Tooth Cutter except as noted.

MATERIAL	CONDITION & MICROSTRUCTURE	BHM	TOOL MATL. TRADE NAME	TOOL MATL. INDUS-TRY GRADE	UP OR DOWN MILL-ING	AR°	RR°	CA°	INCL°	TR°	ECEA°	END REL° COR. REL°	CUTTING FLUID Code **	DEPTH OF CUT in.	WIDTH OF CUT in.	FEED ipt	TOOL LIFE END POINT in.
ALLOY STEELS 4340	NORMALIZED ACICULAR	320	370	C-6	UP	0	-7	45	5	-5	5	6 / 6	00	.100	4.0	.005	.015

TOOL LIFE/TOOTH — inches work travel vs SPEED-feet/minute

(1)	(2)	(3)	(4)
47	70	83	95
670	540	445	345

PERIPHERAL END MILLING

MATERIAL	CONDITION & MICROSTRUCTURE	BHM	TOOL MATL. TRADE NAME	TOOL MATL. INDUS-TRY GRADE	CUTTER TYPE	DIA. in.	NO. TEETH	FLUTE LENGTH in.	UP OR DOWN MILL-ING	HELIX ANGLE°	RR°	ECEA°	CHAM-FER	END REL° PERIPH. REL°	CUT-TING FLUID Code **	DEPTH OF CUT in.	WIDTH OF CUT in.	FEED ipt	TOOL LIFE END POINT in.
ALLOY STEELS 4340	NORMALIZED ACICULAR	331		M2 HSS	SOLID HSS DOUBLE END	.750	4	2	DOWN	30	10	45°	.060	3 / 7	11 1:20 FLOOD	.250	.750	.004	.012

TOOL LIFE/CUTTER — inches work travel vs SPEED-feet/minute

(5)	(6)	(7)	(8)
50	100	140	220
155	100	80	70

END MILL SLOTTING

MATERIAL	CONDITION & MICROSTRUCTURE	BHM	TOOL MATL. TRADE NAME	TOOL MATL. INDUS-TRY GRADE	CUTTER TYPE	DIA. in.	NO. TEETH	FLUTE LENGTH in.	UP OR DOWN MILL-ING	HELIX ANGLE°	RR°	ECEA°	CHAM-FER	END REL° PERIPH. REL°	CUT-TING FLUID Code **	DEPTH OF CUT in.	WIDTH OF CUT in.	FEED ipt	TOOL LIFE END POINT in.
ALLOY STEELS 4340	NORMALIZED ACICULAR	331		M2 HSS	SOLID HSS DOUBLE END	.750	4	2	DOWN	30	10	45 x	.060	3 / 7	11 1:20	.250	.750	.002	.012

TOOL LIFE/CUTTER — inches work travel vs SPEED-feet/minute

(9)	(10)	(11)
50	110	190
65	54	43

DRILLING

MATERIAL	CONDITION & MICROSTRUCTURE	BHM	DRILL MATL. TRADE NAME	DRILL MATL. INDUS-TRY GRADE	TYPE DRILL	DIA. in.	LENGTH in.	FLUTE LENGTH in.	TYPE POINT	HELIX ANGLE°	POINT ANGLE°	LIP RE-LIEF°	CUTTING FLUID Code **	DEPTH OF HOLE in.	FEED ipr	DRILL LIFE END POINT in.
ALLOY STEELS 4340	NORMALIZED ACICULAR	331		M1 HSS	TWIST	.250	2.5	1.375	STANDARD	29	118	7	11 1:20	.5 THRU	.002	.015 / .005

DRILL LIFE — NO. OF HOLES vs SPEED-feet/minute

(12)	(13)	(14)	(15)
90	74	60	150
20	50	50	50

(16)	(17)	(18)	(19)
10	50	45	180
74	110	36	30

** Cutting Fluid Code   00 Dry   11 Soluble Oil

Fig. 4-16.—Tool-life data for face milling, peripheral end milling, end-mill slotting, and drilling a part on an N/C machining center.

N/C MACHINING COSTS AND OPERATION TIMES
FOR
PART NO.—456987-002 * PART NAME-BRACKET SLIDER * LOT SIZE-20
MATERIAL-AISI 4340 STEEL, NORMALIZED, ACICULAR * HARDNESS-331

DATA SET NO.	OPER. NO.	MACHG OPER.	TOOL MATL.	SPEED SFM	FEED	TOOL LIFE	FEED COST $	RAPD TRAV $	LOAD UNLD $	SET UP $	TOOL INDX $	TOOL REPL $	TOOL DEPR $	TOOL SHPN $	BLAD RSET $	INST COST $	GRND WHL $	TOOL PRST $	TOTAL COST $	OPER TIME MIN
30001	N010	FM-TA	C-6	670.	0.005	47.	0.64	0.23			0.08	0.08	0.00			0.06		0.12	1.24	1.27
30002	N010	FM-TA	C-6	540.	0.005	70.	0.80	0.23			0.08	0.05	0.00			0.04		0.08	1.31	1.42
30003	N010	FM-TA	C-6	445.	0.005	83.	0.97	0.23			0.08	0.04	0.00			0.03		0.06	1.45	1.62
30004	N010	FM-TA	C-6	345.	0.005	95.	1.25	0.23			0.08	0.04	0.00			0.03		0.05	1.71	1.96
30005	N020	EM-HSS	M2	155.	0.004	50.	2.78	0.15			0.08	0.32	0.16	0.33			0.01	0.33	4.18	4.02
30006	N020	EM-HSS	M2	100.	0.004	100.	4.31	0.15			0.08	0.16	0.08	0.16			0.00	0.16	5.13	5.67
30007	N020	EM-HSS	M2	80.	0.004	140.	5.38	0.15			0.08	0.11	0.05	0.11			0.00	0.11	6.04	6.91
30008	N020	EM-HSS	M2	70.	0.004	220.	6.15	0.15			0.08	0.07	0.03	0.07			0.00	0.07	6.65	7.79
30009	N030	EM-HSS	M2	65.	0.002	50.	3.33	0.25			0.08	0.10	0.03	0.10			0.00	0.10	4.03	4.55
30010	N030	EM-HSS	M2	54.	0.002	110.	4.01	0.25			0.08	0.04	0.01	0.04			0.00	0.04	4.51	5.30
30011	N030	EM-HSS	M2	43.	0.002	190.	5.04	0.25			0.08	0.02	0.00	0.02			0.00	0.02	5.47	6.52
30012	N040	DRILL	M1	90.	0.002	10.	0.45	0.22			0.08	0.24	0.00	0.08				0.20	1.30	1.21
30013	N040	DRILL	M1	74.	0.002	25.	0.55	0.22			0.08	0.09	0.00	0.03				0.08	1.07	1.15
30014	N040	DRILL	M1	60.	0.002	50.	0.67	0.22			0.08	0.04	0.00	0.01				0.04	1.09	1.25
30015	N040	DRILL	M1	50.	0.002	80.	0.81	0.22			0.08	0.03	0.00	0.01				0.02	1.19	1.39
30016	N040	DRILL	M1	74.	0.005	5.	0.22	0.22			0.08	0.49	0.00	0.16				0.40	1.61	1.23
30017	N040	DRILL	M1	45.	0.005	25.	0.36	0.22			0.08	0.09	0.00	0.03				0.08	0.88	0.92
30018	N040	DRILL	M1	36.	0.005	50.	0.45	0.22			0.08	0.04	0.00	0.01				0.04	0.87	0.97
30019	N040	DRILL	M1	30.	0.005	90.	0.54	0.22			0.08	0.02	0.00	0.00				0.02	0.91	1.06
LOAD, UNLOAD AND SETUP TIME AND COST									1.66	2.49									4.15	5.00

NOTE— FEED UNITS ARE IPR FOR TURNING, DRILLING, REAMING, AND CENTERDRILLING OR CHAMFERING
FEED UNITS ARE IPT FOR MILLING

TOOL LIFE UNITS ARE IN. FOR MILLING, DRILLING, REAMING, AND TAPPING
TOOL LIFE UNITS ARE MIN. FOR TURNING
TOOL LIFE UNITS ARE NO. OF HOLES FOR CENTERDRILLING OR CHAMFERING

N/C MACHINING COSTS AND OPERATION TIMES
FOR
PART NO 456987-002 * PART NAME-SLIDER * LOT SIZE-20
MATERIAL-AISI 4340 STEEL, NORMALIZED, ACICULAR * HARDNESS-331

DATA SET NO.	OPER. NO.	MACHG OPER.	TOOL MATL.	SPEED SFM	FEED	TOOL LIFE	FEED COST $	RAPD TRAV $	LOAD UNLD $	SET UP $	TOOL INDX $	TOOL REPL $	TOOL DEPR $	TOOL SHPN $	BLAD RSET $	INST COST $	GRND WHL $	TOOL PRST $	TOTAL COST $	OPER TIME MIN
30001	N010	FM-TA	C-6	670.	0.005	47.	0.64	0.23			0.08	0.08	0.00			0.06		0.12	1.24	1.27
30005	N020	EM-HSS	M2	155.	0.004	50.	2.78	0.15			0.08	0.32	0.16	0.33			0.01	0.33	4.18	4.02
30009	N030	EM-HSS	M2	65.	0.002	50.	3.33	0.25			0.08	0.10	0.03	0.10			0.00	0.10	4.03	4.55
30018	N040	DRILL	M1	36.	0.005	50.	0.45	0.22			0.08	0.04	0.00	0.01				0.04	0.87	0.97
LOAD, UNLOAD AND SETUP TIME AND COST									1.66	2.49									4.15	5.00
							7.20	0.85	1.66	2.49	0.32	0.54	0.19	0.44	0.06	0.01	0.59		14.47	15.81

NOTE— FEED UNITS ARE IPR FOR TURNING, DRILLING, REAMING, AND CENTERDRILLING OR CHAMFERING
FEED UNITS ARE IPT FOR MILLING

TOOL LIFE UNITS ARE IN. FOR MILLING, DRILLING, REAMING, AND TAPPING
TOOL LIFE UNITS ARE MIN. FOR TURNING
TOOL LIFE UNITS ARE NO. OF HOLES FOR CENTERDRILLING OR CHAMFERING

Fig. 4-17.—Computer printout of N/C machining costs and operation times for Example 2, a part to be machined on an N/C machining center.

30001, gives the lowest cost, $1.24, and the lowest operation time, 1.27 min, of the four cutting speeds investigated. A still higher cutting speed might yield an even lower cost, however, for the feed cost would continue to decrease. However, the tool replacement, insert, and tool presetting costs would increase as shown in the printout.

For peripheral end-milling operation N020, the machining conditions of data set 30005 yield the lowest cost of $4.18 and the lowest operation time of 4.02 min. The data for operation N030, end-mill slotting, shows the lowest cost of $4.03 and the lowest operation time of 4.55 min for data set 30009. These first three operations all show similar results; a minimum cost or minimum operation time has not yet been reached. Projections of the data are in order to determine the cutting speed and tool life for minimum cost and maximum production. The individual cost factors show that as the cutting speed decreases the feed cost increases. However, the tool changing and reconditioning costs decrease. Therefore, the optimum conditions should be calculated and checked by cutting tests.

Two different feed rates are compared in the drilling operation N040. A minimum cost and operation time occurs near the machining conditions given by data set 30013 for a .002-ipr feed rate. However, investigation of a higher feed rate, .005 ipr, shows that a still lower cost, a minimum for that feed rate, occurs for data set 30018. A minimum operation time occurs for the conditions given by data set 30017. Projections of data would not be necessary for these conditions.

By totaling the lowest costs for each operation (data sets 30001, 30005, 30009, and 30018) plus the sum of the handling and prorated setup costs, we find the total part cost for this part.

Operation	Cost
30001	$ 1.24
30005	4.18
30009	4.03
30018	.87
Load, unload, and setup	4.15
Total part cost	$14.47

Adding the corresponding time figures gives a total floor-to-floor time of 15.81 min. Dividing 60 min by 15.81 min yields a production rate of 3.79 pieces per hour.

DETERMINING OPTIMUM CUTTING CONDITIONS

In cases where a mathematical expression relating tool life to cutting speed can be determined from actual tool-life data, it is possible to optimize the general equations and solve for cutting speeds for minimum costs and maximum production rates. This is done by applying Taylor's equation, namely:

For turning—

$$vT^n = S$$

For milling, drilling, reaming, or tapping—

$$vT_t^n = S_t$$

From Equation 4–1, it will be seen that the product of cutting speed (v) and tool life (T) affects the cost of the following quantities: dull tool replacement, tool depreciation, tool resharpening, rebrazing, inserts, grinding wheels, and tool presetting when turning with brazed carbide tools. Some of these items drop out in the case of throw-away inserts and HSS tools, as previously mentioned. Similarly, the tool life in inches affects quantities in Eqs. 4–2 through 4–5 for milling, drilling, reaming, and tapping.

By substituting Taylor's equation into the cost and operation time equations, we derive a new series of equations to determine the cutting speed and tool life for minimum cost and for maximum production. The cutting speeds for these conditions are given in Eqs. 4–13 and 4–14 for turning, Eqs. 4–15 and 4–16 for milling, Eqs. 4–17 and 4–18 for drilling or reaming, and Eqs. 4–19 and 4–20 for tapping, as shown in Table IV–4. The tool life in minutes corresponding to minimum cost and maximum production is given in Eqs. 4–21 and 4–22 for turning and Eqs. 4–23 and 4–24 for milling, drilling, reaming, or tapping, as shown in Table IV–5. The terms $v_{min.\ cost}$ and $v_{max.prod.}$ must be calculated using the appropriate equation corresponding to the proper operation from Eqs. 4–13 through 4–20. For some of the various types of tools used in each operation, the nonapplicable terms will drop out.

Table IV-4. Optimized Cutting Speed Equations for Turning, Milling, Drilling, Reaming, and Tapping.

Turning

$$v_{min.\,cost} = \left[\frac{n\,M\,(L+e)}{L\,(1-n)\,\left(M\,t_d + \dfrac{C_p}{k_1+1} + G\,t_s + \dfrac{G\,t_b}{k_2} + \dfrac{C_c}{k_3} + C_w + G\,t_p \right)} \right]^n S \qquad (4\text{--}13)$$

$$v_{max.\,prod.} = \left[\frac{n\,(L+e)}{(1-n)\,L\,t_d} \right]^n S \qquad (4\text{--}14)$$

Milling

$$v_{min.\,cost} = \left[\frac{n\,M\,D\,(L+e)}{3.82\,f_t L\,\left(M\,t_d + \dfrac{C_p}{k_1+1} + G\,t_s + \dfrac{G\,t_b}{k_2} + \dfrac{Z\,C_c}{k_3} + C_w + G\,t_p \right)} \right]^{n/(n+1)} S_t^{1/(n+1)} \qquad (4\text{--}15)$$

$$v_{max.\,prod.} = \left[\frac{n\,D\,(L+e)}{3.82\,f_t L\,t_d} \right]^{n/(n+1)} S_t^{1/(n+1)} \qquad (4\text{--}16)$$

Drilling or Reaming

$$v_{min.\,cost} = \left[\frac{n\,M\,D\,(L+e)}{3.82\,f_r L\,\left(M\,t_d + \dfrac{C_p}{k_1+1} + G\,t_s + G\,t_p \right)} \right]^{n/(n+1)} S_t^{1/(n+1)} \qquad (4\text{--}17)$$

$$v_{max.\,prod.} = \left[\frac{n\,D\,(L+e)}{3.82\,f_r L\,t_d} \right]^{n/(n+1)} S_t^{1/(n+1)} \qquad (4\text{--}18)$$

Tapping

$$v_{min.\,cost} = \left[\frac{m\,n\,M\,D\,(L+e)}{1.91\,L\,\left(M\,t_d + \dfrac{C_p}{k_1+1} + G\,t_s + G\,t_p \right)} \right]^{n/(n+1)} S_t^{1/(n+1)} \qquad (4\text{--}19)$$

$$v_{max.\,prod.} = \left[\frac{m\,n\,D\,(L+e)}{1.91\,L\,t_d} \right]^{n/(n+1)} S_t^{1/(n+1)} \qquad (4\text{--}20)$$

Table IV-5. Optimized Tool Life Equations for Turning, Milling, Drilling, Reaming, and Tapping.

Turning

$$T_{min.\,cost} = \left[\frac{S}{v_{min.\,cost}} \right]^{1/n} \qquad (4\text{--}21)$$

$$T_{max.\,prod.} = \left[\frac{S}{v_{max.\,prod.}} \right]^{1/n} \qquad (4\text{--}22)$$

Milling, Drilling, Reaming, or Tapping

$$T_{t\,min.\,cost} = \left[\frac{S_t}{v_{min.\,cost}} \right]^{1/n} \qquad (4\text{--}23)$$

$$T_{t\,max.\,prod.} = \left[\frac{S_t}{v_{max.\,prod.}} \right]^{1/n} \qquad (4\text{--}24)$$

Minimum Cost and Maximum Production

Substituting the calculated optimum cutting-speed and tool-life values for minimum cost into their respective machining cost equations, Eqs. 4–1 through 4–5, will give the minimum cost for each operation. Eq. 4–6 should also be solved for the handling and setup costs and should be added to the operation costs to give the minimum total cost per piece in machining for all the operations performed on the N/C machine tool.

By calculating the optimum cutting speed and tool life for maximum production and substituting these values into the appropriate operation time per piece from Eqs. 4–7 through 4–11, the minimum operation times can be found which are then used to determine the maximum production rate as described by Eq. 4–12.

Restraints

To determine whether Taylor's equation is applicable in any given machining operation, tool life vs. cutting speed can be plotted on bilogarithmic paper and can be observed for a straight-line relationship as described in Chapter 3. One can comfortably interpolate between data points on this log-log plot if a straight line is obtained. However, extensive extrapolation of data beyond the experimental values of tool life may lead to erroneous conclusions unless the data are verified in the shop.

Taylor's equation applies to about 75 to 80 percent of turning operations. It gives a straight line on log-log graph paper and can be used conveniently in cost and production analyses. In about 25 percent of the cases, a curve will result instead of a straight line. If a straight logarithmic line is not obtained for the T-v relationship from experimental data, it can often, but not always, be concluded from dimensional analysis that temperature conditions have changed (9). When the T-v relationship is not a straight line on log-log paper, Taylor's equation can be replaced by a quadratic or similar equation. The application of this type of equation, however, is more complex than the application of Taylor's equation and will not be considered here.

There is a series of practical restraints which must be met in any machining operation; accuracy, surface finish, and surface integrity required by the component must be satisfied. Cutting forces or torque should not exceed the breaking point of the tools and cutters involved. Generally, tool life must be greater than feed time for one pass over the workpiece. If this is not the case, the tool will break down in the middle of the cut, requiring a costly interruption of the machining operation. Such interruptions are also highly undesirable from the standpoint of accuracy, because it is hardly possible to continue a cut after replacing a tool without leaving a mark on the workpiece. In addition, the machine control tape might not be prepared to continue from the point of tool breakdown, requiring a costly loss of time in recycling the tape from the beginning of the operation.

The horsepower required for the cut must be calculated to make sure it does not exceed the horsepower available at the spindle. Similarly, the required spindle speed must be compared to the spindle speeds available. When large differences occur, the effects on tool life, cost, and production rate must be determined.

DEVELOPING A TOTAL SYSTEM

The first step in developing a total cost and production analysis system is to collect data for each material in use in the form shown in Figs. 4–1 through 4–5 and as discussed at the beginning of this chapter (8). In addition, the unit horsepower (hp/cu in./min) is required data for each material. This machinability data is then used to build a *material file* for random access as required.

A *machine-tool file* is necessary to record all the relevant information for each N/C machine tool. This file should include the machine identification and description, available speeds and feeds, the horsepower available at each speed, physical dimension constraints, and other data pertinent to each N/C machine tool.

A *time-study and cost file* should contain information relative to the tool costs shown in Figs. 4–12 and 4–15. This file should also include labor and overhead rates, tool reconditioning costs, and required time-study data.

The part geometry and the process description should be entered into the computer program. The program will make the calculations and print the results as shown in Figs. 4–10 and 4–17, having retrieved the cutting parameters from the *material file* and the relevant data from the *time-study and cost file*. A check should then be made to determine the availability of the feeds and speeds required for a specific machine tool, and the nearest available feeds and speeds should be selected. When optimum conditions are desired, a test could be made to see if Taylor's equation will hold (i.e., whether the data will plot as a straight line on log-log paper). If so, then optimum conditions might be determined from Eqs. 4–13 through 4–24.

REFERENCES

1. Inyong Ham, "Economics of Machining: Analyzing Optimum Machining Conditions by Computers," *ASTME Paper No. MR64-534* (1964).
2. C. E. Downing, "Setting Machining Feeds and Speeds: Optimization and Machining Economics," *ASTME Paper No. MR62-152* (1962).
3. N. N. Zorev, "The Effect of Tool Wear on Tool Life and Cutting Speed," Russian Engineering Journal, Vol. XLV, No. 2 (1965), pp. 63–70.
4. Hans Ernst and Michael Field, "Speed and Feed Selection in Carbide Milling With Respect to Production, Cost, and Accuracy," *Transactions of the ASME,* Vol. 68, April (1946), pp. 207–15.
5. K. M. Gardiner, "Computer Decides Conditions for Minimum Cost Machining," *Metalworking Production,* Vol. 109, No. 49 (December 8, 1965), pp. 61–64.
6. E. J. Weller and C. A. Reitz, "Optimizing Machinability Parameters with a Computer," *ASTME Paper No. MS66-179* (1966).
7. Michael Field, Norman Zlatin, Roy L. Williams, and Max Kronenberg, "Computerized Determination and Analysis of Cost and Production Rates for Machining Operations, Part I—Turning," *Transactions of the ASME,* Vol. 90, August (1968), pp. 455–66.
8. Michael Field, Clarence Mehl, and John F. Kahles, *Machining Data for Numerical Control, (AFMDC 66-1).* Cincinnati: Air Force Machinability Data Center, 1966.
9. Max Kronenberg, *Machining Science and Application.* New York: Pergamon Press, 1966, p. 57.

FUTURE MACHINABILITY DATA CONCEPTS

Charles F. Carter, Jr., Cincinnati Milacron, Inc.

As has been stated in earlier chapters of this book, the machinability data used for numerical control does not differ greatly from that used for conventional machining. It has been implied that numerical control merely makes the wise use of good data more important to economical production. This same general statement will be even more relevant for future systems, because they will be increasingly able to use stored machinability data with a minimum of human intervention.

ROLE OF THE COMPUTER

The most important impact of machinability data in the future will be created by computers. With their ability to store and manipulate vast amounts of information, computers will allow advance planning of many factors in machining operations which must now be worked out on the shop floor.

Machining Cost Elements

Historically, the floor-to-floor cost incurred by a workpiece on a given machining operation has been broken down into a number of elements as shown in the following equation:

$$C_T = C_m + C_s + C_l + C_t \tag{5-1}$$

Where:

C_T = Total cost of the piece
C_m = Machining cost per piece
C_s = Cost of setup amortized over the number of pieces
C_l = Cost of loading and unloading an individual piece
C_t = Tool cost per piece.

In order to act on an analysis of this cost equation, the user must know each of the cost elements with reasonable accuracy. This knowledge in turn involves the knowledge of such factors as (1) the setup time, (2) the tool-change time, (3) the actual ma-

157

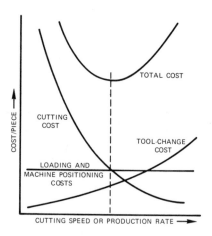

Fig. 5-1.—Machining costs as functions of cost per piece vs. cutting speed or production rate.

chining time, and (4) the labor and burden rate assigned to different operations such as setup, cutter grinding, and machining. These factors are frequently not well known because of the tremendous task of gathering and correlating all of the data necessary to find their accurate values. As a result, the set of curves shown in Fig. 5-1 are either not used in average shop procedures or are used in a generalized form because of uncertainty. Because computers can now be used as tools to gather and store cost and time data on shop operations, the concept of finding that machine operating point at which minimum cost is incurred will be used increasingly in the future.

Data Needs

Increased use of this concept will increase the need for better machinability data. Machinability data is the primary ingredient in the cutting-cost curve shown in Fig. 5-1. Traditionally, the shape of this curve has been determined by using Taylor's equation, $VT^n = C$. This expression assumes that tool life, T, is primarily a function of the cutting speed, V. Practice has shown that the values of the exponent n can only be relied upon if they are derived from a metal-cutting situation very much like the one for which the tool life is being calculated. In other words, without extensive machinability data on many operations, ability to predict tool life is seriously hampered.

One of the reasons for the inadequacy of Taylor's equation is that it attempts to lump together many factors which influence tool life, such as workpiece hardness, rigidity of the cutting system, and chip thickness. As shown in earlier chapters, tool geometry conditions can also have an important effect on tool life. All this emphasizes the need to be able to store a mass of cutting data from many different situations and extrapolate from the mass only that data which is important to a given situation. The computer is the tool by which this can be done.

Present Computer Systems

At this time, one of the most advanced systems for using machinability data in conjunction with a computer is incorporated in the EXAPT part programming system (1).

In the EXAPT system, as described in Chapter 3, technological information with respect to the cutting process is stored in the computer and is used to select machining conditions for the parts programmer. For example, for drilling, a complete file of tool descriptions is stored in the computer. This requires the assignment of a numerical code to drills according to all combinations of length, diameter, shank type, and other characteristics. In addition, the system's files contain a list of tool-life curves specifically for drilling operations and related to material. These curves may be provided by the EXAPT management, in which case they represent a correlation of industry-wide practice, or they may be developed by the individual user.

In using the EXAPT system, the parts programmer merely specifies the metal-cutting operation— DRILL, for example—and in addition indicates the depth and diameter. The computer then selects the proper tool and calculates the correct feed rate and speed. In a further sophistication of this system, a file in the computer also contains machine-tool characteristics. Once the machine tool has been selected by the programmer, the computer compares the cutting requirements with the capabilities of the machine and alters feed-rate and speed selections to keep within torque and spindle speed range capabilities.

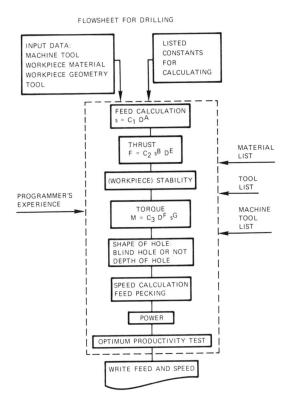

FLOWSHEET FOR DRILLING

Fig. 5–2.—Flow chart for a drilling operation with a computerized machinability data system (2).

Other systems for computer-aided selection of feed and speed have been suggested (2), and a typical flow chart for a drilling operation with one of these systems is shown in Fig. 5–2. Note its similarity to the EXAPT system's philosophy. The inputs are essentially the same, and the considerations for feed thrust and torque are also included. The Optimum Productivity Test block near the end of the flow chart suggests a sophistication which would require the storage of considerable cost or economic data in addition to machinability data.

A similar philosophy is used by the EXAPT system to aid in the selection of the variables for turning. By comparing machine-tool power limitations, tool geometry, finished size, and surface finish, the computer will select maximum allowable depth of cut, maximum feed rate (where tool geometry and surface finish would be considered for a finish cut), and cutting speed, which also uses a tool-life constraint in addition to horsepower.

Future Computer Systems

Many other systems which will store and manipulate machinability data have already been developed, and others undoubtedly will be developed in the future. One of the advances which future systems will probably incorporate will be the ability to change stored data easily. Data provided by a central source must necessarily be generalized, but many users have special conditions which make it desirable for them to modify generalized machinability data. Therefore, the full value of the computer will not be realized until the part programming languages which are based on machinability data are written so that the data bases may be easily altered. The bases may be safely altered only by those users who have found data-retrieval methods for machinability information.

OPTIMIZING THE CUTTING PROCESS

Although the most sophisticated systems may be developed for selecting proper machining conditions, it is unlikely that ideal machining conditions can be used unless speeds and feeds can be changed at the machine tool under actual operating conditions. This hesitancy to use optimum cutting conditions results because variations in workpiece hardness and size, coupled with variable tool conditions, can only be accounted for in advance by statistical grouping. In order to optimize the conditions under which each individual part is made, the conditions must be changed at the machine. A partial solution to this problem is provided by direct numerical control (DNC). The complete solution will probably be provided by some form of adaptive control.

Direct Numerical Control (DNC)

The term *direct numerical control*, like *adaptive control*, is fast becoming a broadly used term which describes systems of slightly different configuration but similar purpose. In order to understand the impact of DNC and, more specifically, its relationship to machinability data, it is first necessary to fully understand the flow of information in a conventional N/C system as shown in Fig. 5–3.

Conventional Control. In order to make a useful tape for an N/C machine, workpiece geometry data must be derived from the part drawing, either manually or with the aid of some program such as APT, as discussed in Chapter 2. Machining conditions are then added to this geometry data, and this package of information is

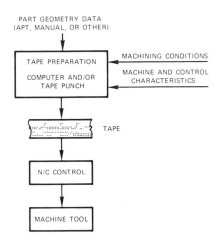

PART GEOMETRY DATA
(APT, MANUAL, OR OTHER)

TAPE PREPARATION

COMPUTER AND/OR
TAPE PUNCH

MACHINING CONDITIONS

MACHINE AND CONTROL
CHARACTERISTICS

TAPE

N/C CONTROL

MACHINE TOOL

Fig. 5-3.—Simplified flow chart of a conventional N/C operation sequence (2).

processed in order to make it acceptable to the particular machine and control combination on which the tape will be used. Usually, when the tape is first tried on the machine, changes are required in order to make a successful part. This revision may be necessary in spite of the fact that the geometry data may have been previously checked on some type of plotting device. Actual running of the tape reveals such problems as errors in speed and feed selection, interference with workholding fixtures, and the need for programmed stops to clear chips. The mechanics of adding these changes to the tape can require hours or even days. Valuable production time is lost while the tape is changed, and, due to the awkwardness of the tape-changing procedure, many changes which would improve productivity are not made once a successful part is being produced.

Conventional Management. Under conventional N/C, management's link with the machine tool is dependent on the same reporting systems which have been used for decades. For this reason, such important factors as machine utilization and work-in-progress are presented to management somewhat inaccurately and with considerable time lag. This slowness and inaccuracy results because methods used to provide management with information usually involve the manual collection, the manual or automatic transmission and storage, and the manual manipulation of data for useful presentation. Management action cannot be taken on a real-time basis because facts are not known on a real-time basis.

Advantages of DNC. The two major advantages of DNC over conventional N/C are the provision of means for (1) real-time program editing and (2) handling of information necessary for management control. Fig. 5–4 depicts the essential elements of a DNC system. The data normally put on tape is acquired in the same manner as for conventional N/C; however, it is stored in some device such as a magnetic disk. A computer, which may be called a "central data source" or "director," acts primarily to control and supervise the flow of information to and from the various devices which will make up the complete system. In the example shown in Fig. 5–4, the program for a given part may be called for from the machine. The central data source will select

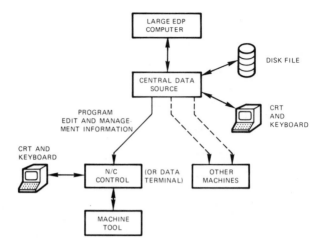

Fig. 5-4.—Flow chart of direct numerical control (DNC) operations (2).

information from the disk file and send it to the appropriate machine and control as it is needed. The link to the large computer would be used to handle management or supervisory information and to load the local disk file from some master file or to make corrections in the part program in the basic language, such as APT, when such corrections are required.

When a part is being run for the first time and errors are discovered or clearances should be checked in advance, machine control information may be displayed on the cathode ray tube (CRT) at the machine. A keyboard is provided from which changes in the data may be made and new conditions tried. Since the procedure for changing data is simple and instantaneous, even minor modifications which will improve the operation may be made. Once changes have been proven, they can automatically become a part of the permanent program in the disk file. Through the use of sensors on the machine or keyed inputs from the operator, information vital to management can be updated continuously and accurately. Such information might include part number and process, cycle time, horsepower consumed, tool life, and many other similar management tools.

Data for DNC. The role that machinability data will play in such a system is very great for two reasons:

1) Machinability data can be expected to be much more accurate once the full management control system is working with some system for reporting cutting-tool use.

2) The updated machinability information can be quickly incorporated into the part program, since lengthy and expensive punching of new tapes will not be required.

With an expanded system for the collection, storage, and retrieval of machining data, it is possible that data for particular combinations of workpieces (or classes of workpieces) and machines can be made available, and, from this, much more accurate operating points can be established. The introduction of new cutting-tool materials or changes in workpiece materials which require changes in speed or feed can be accommodated quickly through the use of the CRT and the keyboard. The changes can be made at the machine when the part is run, or revisions can be made "off line" by calling for the part information to be displayed on an appropriate CRT and by making changes through an associated keyboard.

Adaptive Control

Fig. 5–5 shows a simplified block diagram of an adaptive control system. Since feed rate and spindle speed are the only two machining variables which can be changed on line, these are the "machining rate modification commands" which are sent from the adaptive controller to the N/C control unit. The sensor signals—which may be torque, deflection, temperature, vibration, and others—are sent to the adaptive controller. The adaptive controller acts on these signals according to a predetermined reaction to the machining variables, or "strategy," (in this case) and from them generates an output to the N/C control unit. The signal to the N/C control unit acts to modify the speed or feed command on the control tape. The signals from the N/C control to the machine are handled in the conventional manner, but are continually modified by the adaptive controller rather than being the exact image of the control-tape input.

Data for Adaptive Control. Fig. 5–6 shows the influence of adaptive control on some common process variables. Under operating conditions in which variable width or depth of cut will be encountered, the programmer must select machining conditions so that the part will be in tolerance under the worst conditions. As Fig. 5–6 indicates, the adaptive rate will slow down to the most conservative value only at the most critical point. As tools wear, deflections increase, and depth of cut must be chosen to allow for machining under the worst tool-wear conditions. The adaptive control system senses the effect of tool wear and slows down only as wear increases. Increased deflection brought about by increased hardness can also be sensed by the system, and proper tolerance can be maintained in spite of variable hardness. Finally, considerable improvements in productivity can be gained by speeding up over air gaps which the programmer may overlook unless the tape is completely optimized manually.

The use of adaptive control greatly reduces the requirement for accurate selection of speeds and feeds during the planning stage. However, the need for accurate machina-

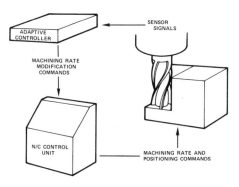

Fig. 5–5.—Relationship of an adaptive controller to an N/C control unit and machine tool.

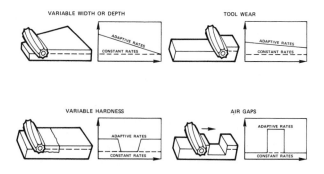

Fig. 5–6.—Optimization of production by reduction of the effects of common process variables with adaptive control.

bility data will continue to exist, since it must be used for the proper selection of machine tools, cutting tools, fixturing, and broad feed and speed ranges. In order to take advantage of adaptive control, the parts programmer will need data which he has not used in programming with conventional control. He will need cutter and machine deflection data, since he can program for a desired deflection or for the maximum strength of the cutter or machine. Since forces and torque on the cutter can be accurately controlled, those limits on the cutter must also be known. Chip loads which are considered safe maximums or minimums for given cutter/work-material combinations should be known if these factors become constraints before deflection or the machine's limits are reached. In short, adaptive control will provide optimum instantaneous operating conditions, but the range, or bounds, of operation must be set by the programmer based on accurate data concerning the machine, cutting tool, and work material.

Types of Adaptive Control. Since adaptive control is such a broad term, it is necessary to review the characteristics of some systems which may be referred to as adaptive control systems. The simple examples below show these characteristics in order of increasing complexity.

Simple Feedback Control. A simple feedback control system is shown in Fig. 5–7. In this example, the command may be a feed rate or a slide velocity. The object of the control is to make the machine-tool slide move at the commanded velocity regardless of outside conditions. The feedback element senses the slide velocity and compares it to the commanded velocity or feed rate. In the case of the simple feedback control, the control is independent of and unresponsive to any process variables. For this reason, it is not an actual adaptive control system; however, it does show how the feedback principle is used in more advanced systems.

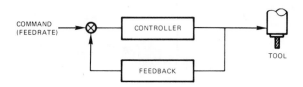

Fig. 5–7.—Simple feedback type of adaptive control system.

Adaptive Constraint Control. Fig. 5–8 shows a system in which another loop has been added to the control. In this case, a process variable is measured and the initial command is altered depending on the condition of the process variable. This control system is called *constraint control* because some limiting value, or constraint, is usually selected for the process variable, which may be torque, deflection (force), or horsepower. This system is the simplest form of adaptive control, and when the process variables which will be used as constraints are properly sensed, it can be very effective.

Programmed Adaptive Control. In order to discuss more sophisticated forms of adaptive control, a new term called *performance index* must be introduced. The performance index is the relationship of all process variables which leads to optimum performance. The index is usually expressed in terms of cost or production rate, but it could also be expressed in terms of accuracy or surface finish if these represent the most demanding or important products of the process.

Much has been written about the relationship between cost, production rate, and tool life; however, the mathematical models found in metal-cutting literature are sel-

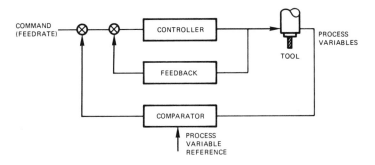

Fig. 5–8.—Adaptive constraint control system.

dom used in practice. The reason for the infrequent use of these models probably lies in the fact that the constants which must be fed into them are not known to any great degree of accuracy, nor are they easy to obtain. Computer-aided methods of data manipulation, coupled with computer-aided process monitoring methods, will allow these models to become more widely used. These models, and the general philosophy on which they are based, are important in deriving the performance index for an adaptive control system.

Fig. 5–9 shows the block diagram of a programmed adaptive control system in which a performance index is used. Here, the process variables are measured as in adaptive constraint control, but the system's reaction to the process variables depends on a performance index rather than on simple limits for the variables. In this case, the performance index is calculated before it is entered into the system, and a strategy is predetermined.

One system in use today falls into this category (3). In this system, the performance index is cost based on tool life. Cost for tool life is calculated before it is programmed into the system, so the direct performance index for the system is actually tool life. A fixed reaction to the index then responds to the process variables of torque and deflection which act to drive the system to achieve the desired tool life.

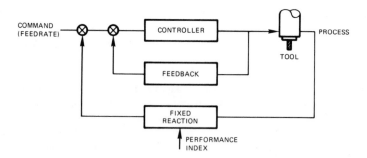

Fig. 5-9.—Programmed adaptive control system.

Optimal Adaptive Control. Fig. 5–10 represents an even more sophisticated form of adaptive control. In this system, the mathematical model of the performance index is included in the control, and the controller continually calculates the index. The reaction of the system to the calculated index is to attempt to maximize or minimize it. Much has been written about these techniques of optimizing the performance index, and most written discussion of adaptive control centers around the selection of a mathematical "hill-climbing" technique. However, the real problem of this type of control is to measure enough process variables on an instantaneous basis in order to satisfy the mathematical model of the performance index. At present, this measurement problem has not been solved, and all adaptive control systems operating on machine tools today are either of the programmed adaptive type or adaptive constraint type.

NEW USES OF MACHINABILITY DATA

Because the N/C machining cost is usually higher than the cost in a conventional situation, adherence to good N/C machining practice and operation at near optimum points carries with it much more economic significance. Fig. 5–1 shows that machine handling costs (such as slide positioning) and tool-changing costs are an important part

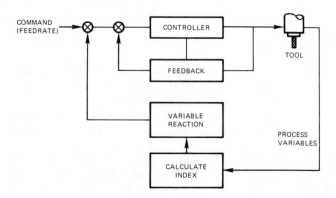

Fig. 5-10.—Optimal adaptive control system.

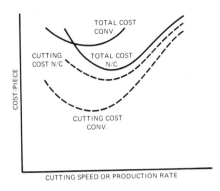

Fig. 5-11.—Comparative machining costs for N/C and conventional machining as functions of cost per piece vs. cutting speed or production rate.

of total costs. The influence of changing these costs can be seen in Fig. 5-11 (2). Note in Fig. 5-11 that even though the machine and cutter cost during actual cutting time is higher for an N/C machine, the total cost is lower because of the influence of reduced positioning and tool-changing time. The most significant factor is that the reduced cost is achieved at a higher cutting speed than for the conventional operation. From this curve we should also recognize that when a tool-changing N/C machine is compared to a conventional N/C machine, the minimum cost operating point will be shifted to a higher cutting speed. The point here is that the speed arrived at as optimum for one machine may not be the correct speed for another machine, even though the operations are identical.

To further complicate the issue, it is very difficult to predict a speed that will produce a given tool life unless a considerable amount of data is known about the particular tool/material combination and the characteristics of the machine. This all emphasizes the importance of gathering and storing accurate machinability data.

Another concept which demonstrates the importance of machinability data is that of automated small-lot manufacturing. These small-lot systems include part handling and machining for a variety of workpieces with complete computer control of the whole system. At the present time, such systems are just beginning to emerge. However, they all tend to reduce the time a particular part is in process and increase the proportion of time (not the absolute time) that a part spends undergoing actual metal removal. Most of the time a part requires for manufacture is spent in waiting for the next operation. Automated systems will greatly reduce this waiting time. The increased relative importance of machining will throw further light on the necessity of running at optimum speeds and feeds. This in turn will require better machinability data aimed at each particular operation.

REFERENCES

1. Dieter Reckziegel, "N.C. Programming System EXAPT," *Proceedings of 1967 Seminar on Horizons in Manufacturing Technology*, University of Michigan, Institute of Science and Technology, Industrial Development Division, April, 1967.

2. Bertil Colding, "Machining Economics and Industrial Data Manuals," paper presented at the CIRP Annual Meeting, University of Manchester, Institute of Science and Technology, Manchester, England, Sept., 1968.

3. R. A. Mathias, "Adaptive Control of the Milling Process," paper presented at the IEEE National Machine Tools Industry Conference, Cleveland, Ohio, Oct., 1967.

CHAPTER 6

MACHINABILITY DATA: AVAILABILITY AND ECONOMICS

John F. Kahles, Metcut Research Associates, Inc.
Norman Zlatin, Metcut Research Associates, Inc.

Throughout this book, the importance of reliable, accurate machinability data to N/C machining has been emphasized. The dependence of efficient N/C operations on such data is reflected in the money and effort that are spent to obtain the most up-to-date, effective machining data and data systems available. Therefore, a knowledge of the overall economics of the metal-cutting industry and the general availability of machinability data for N/C is an important aspect of efficient N/C operations.

MATERIAL-REMOVAL ECONOMY

Material removal is recognized as the most widely used metalworking manufacturing operation in modern industry. Although a highly detailed assessment of the economic importance of machining has really never been made, a plan for making reliable estimates for management purposes was developed by Hans Ernst in 1958 (1). Ernst calculated that the cost of labor and overhead alone at that time amounted to annual expenditures on the order of $14 billion. Additional calculations, as shown in Table VI-1, indicated that over 17 million tons of metal were converted into chips each year. Note the great extent of these expenditures even though Ernst's calculations excluded 600,000 small chip makers such as grinding, honing, lapping, and filing machinery.

In 1968 *The Tenth American Machinist Inventory of Metalworking Equipment* showed that 2,475,000 machine tools were being used in American industry—excluding machines used in training and those in surplus stores. Since labor and overhead rates also apply to small chip makers, a revision of Ernst's calculations shows that approximately $40 billion were spent in labor and overhead in 1968 (2). Cost figures such as those shown in Table VI-2 also justify management's interest in machinability and machining data. Note that machine tool shipments passed the $1 billion level in 1968 (3), and cutting tool shipments by the same year amounted to over $739 million per year.

Additional data in Table VI-3 show the dollar values of individual types of cutting tools, including high-speed steels and carbides. Cutting tools such as carbides and

Table VI-1. Estimate of United States Metal-Cutting Operations, 1958 (1).

Total Number of Machine Tools Installed*		
All categories		2,000,000
Small machines†		600,00
Balance (large machines)		1,400,000
Estimated Weight of Metal Removed		
Average power actually used in metal cutting (hp)	2	
Average efficiency (hp/cu in./min)	2	
Average volume of metal removed per machine (cu in./hr) $= \dfrac{2 \times 60}{2} =$	60	
Average metal cutting time/day (hrs)	6	
Average weight/cu in. metal	.28	
Average working days/yr	250	
Total weight of metal cut each year (tons) $= \dfrac{60 \times 6 \times 250 \times 1{,}400{,}000 \times .28}{2{,}000} =$		17,640,000
Estimated Cost		
Average labor and overhead rate (per hr)	$5.00	
Average working time/day (hrs)	8	
Average working days/yr	250	
Total labor and overhead = $5.00 × 8 × 250 × 1,400,000 =		$14,000,000,000

*From *American Machinist Inventory of Metalworking Equipment*, 1953.
†Grinding, honing, lapping, polishing, filing and other small equipment.

ceramics must be regarded as having inherently better material removal capability, and these materials are more resistant to abrasion and to the high temperatures developed during chip making. However, carbides and, in particular, ceramics lack the toughness characteristics shown by HSS tools. Therefore, HSS tools continue to find highly significant markets.

Goals

Fig. 6–1 sets forth interesting goals for marketing considerations. The total dollars spent for shipments of HSS tools is a significant market area that should be protected by programs to improve the cutting capability of high-speed steels without sacrificing their toughness. At the same time, the current expenditures for high-speed steels pre-

Table VI-2. Metal-Cutting Statistics for the United States, 1968 (2, 3, 4).

Annual Labor and Overhead Costs for Machine Tool Operation		
Total number of machine tools installed		2,500,000
Average labor and overhead rate (per hr)	$8.00	
Average working time/day (hrs)	8	
Average working days/yr	250	
Average number of direct labor personnel per machine	1	
Total labor and overhead = 2,500,000 × $8.00 × 8 × 250 × 1 =		$40,000,000,000
Total Shipments (Including Exports) of Metal-Cutting Machinery		$1,080,000,000
Machine Tool Accessories Industry		
Small cutting tools for machine tools		$693,500,000
Toolholders		$45,500,000

Table VI-3. Values of Shipments of Metal-Cutting Tools by Class
of Cutting Tool and Tool Material, 1968 (4).

Type of Cutting Tool	High Speed Steel	Carbide	Carbon Steel	Cast Alloy, Ceramic, and Diamond
Broaches	$ 33,917,000	$ 1,871,000		
Twist Drills	119,519,000	10,770,000	$ 1,406,000	
Gun Drills and Gun Reamers	1,046,000	5,059,000		
Spade Drills	1,678,000	765,000		
Combination Drills and Countersinks	5,066,000	416,000	(included with HSS)	
Countersinks	3,918,000	1,211,000		
Counterbores	8,393,000	2,721,000		
Reamers	21,890,000	8,086,000	2,868,000	
Hobs	14,775,000			
Gear-Shaper Cutters	8,158,000			
Gear-Shaving Cutters	4,837,000			
End Mills	42,700,000	8,911,000		
Replaceable Inserted Blade Cutters				
Nonindexable Type	7,087,000	7,148,000		
Indexable-Insert Type	1,745,000	6,644,000		
Form-Relieved Cutters	6,443,000	465,000		
Slitting Saws and Screw Slotting Cutters	3,677,000	1,480,000		
Other Milling Cutters	25,611,000	3,045,000		
Taps	64,851,000	500,000	3,078,000	
Threading Dies	1,930,000	249,000	3,664,000	
Threading Sets	2,501,000			
Chasers and Blades for Taps and Dies	15,351,000	471,000		
Single-Point Tools	18,763,000	32,339,000		$4,609,000
Circular Form Tools	3,974,000	4,563,000		
Blanks and Tips		43,763,000		57,000
Inserts, Indexable Types	2,006,000	85,000,000		1,602,000
Other Inserts	2,043,000	17,839,000		1,206,000
Rotary Burrs and Files	2,381,000	3,910,000		
Other Tools Not Listed	5,171,000			
Thread Rolling Dies	13,744,000			
Toolholders for Indexable Inserts	30,375,000			
Total Each Material	$439,175,000	$247,226,000	$11,016,000	$7,474,000
Total Value	$738,906,000			

sent a challenge for research and development to increase applications of modern carbides and ceramics by modification and innovation. Additional economic data could be cited, but certainly these financial data should be sufficient to attract the attention of progressive management.

Problems

Historically, certain characteristics of the machining industry—existing even to the present day in many plants—work against good data management. In earlier days, the machining industry took an artisan approach to machining, with control of removal

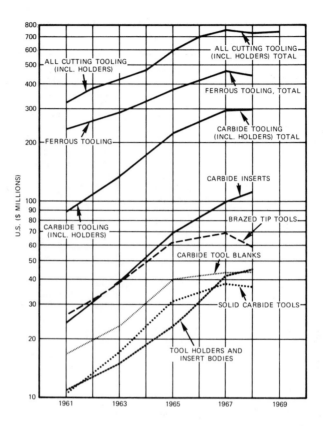

Fig. 6-1.—Growth in carbide tool shipments from 1961 through 1969 (4).

processes manifested primarily in the machine-tool operator. The selection of machining parameters, including feeds, speeds, tool materials, tool geometry, and cutting fluids, rested with the operator and sometimes with the influence of salesmen selling their products. An additional factor which has promoted managerial neglect is that the highly technical and economic aspects of machining are easily overlooked by management because material-removal processes tend to appear disarmingly simple. At present, however, manufacturers are beginning to recognize the importance of machining data in the selection of machine tools, cutting tools, and cutting fluids, and in the influence of work-material choices and even design considerations. Fortunately, an increasing number of managers are realizing the importance of machining data as it contributes to their manufacturing processing and to reduction of costs.

THE MACHINABILITY INDEX

The quality of machining data used in manufacturing varies widely, depending upon production requirements. For many years, considerable emphasis has been placed upon use of the so-called machinability index based on the machining of B1112 steel at 180 sfpm with HSS tooling. In spite of the many disadvantages of this index, including the fact that its values for carbide tools are not proportional to or transferrable from HSS tool data, some organizations have developed rather complex machinability indexes for their various material-removal requirements and are still continuing to use those indexes. Index variations, for example, are based upon the use of cobalt-free HSS tools. If an 8 percent cobalt steel is used, the cutting speed may be increased by a value of approximately 120 percent. Brazed carbides may raise the speed by 250 percent, while carbide inserts may provide values on the order of 300 percent. All operations are related to a basic material-removal operation such as turning. When other operations are used, the speed values are again related by using correction values. Typical operations and the percentages suggested for them are listed as follows:

Operation	Percent of Turning Speed
Spiral-Flute Drilling	50
Tapping	10
Face Milling	55
End Milling	45
Broaching	10

N/C users should seriously consider eliminating machinability indexes from their machining data requirements. The rough approximations of machinability provided by the index can no longer be justified in view of the high cost of machining, including the large capital expenditures for both conventional and N/C equipment. Another limitation of the indexes is that they provide relative speeds only, and they give no information on other significant parameters such as feed, tool geometry, and cutting fluid.

MACHINABILITY TESTS

A number of different methods have been developed which quickly determine the machinability of metals and which include procedures for obtaining a fundamental knowledge of chip formation during metal cutting. The following are a few commonly used testing methods:

1) Tool dynamometers are used to measure the forces acting on the cutting tool.
2) Thermocouples formed between the work and the tool measure tool temperatures.
3) Accelerated tool-life tests can be made using Taylor's equation.
4) Constant-pressure lathes and drills are used to measure the penetration rate of a tool passing through a piece of metal.

Each of these methods is discussed below. A shortcoming of all of the methods is that the tests are usually performed for either turning or drilling, and the metal is then given a rating which is used to develop the machining conditions for any machining operation. No consideration is given to the type of tool, the tool geometry, the cutting fluid, or any of the other factors that might be involved in the given machining operation. For example, data obtained in turning or drilling cannot be used with confidence in setting up the conditions for tapping. Many other shortcuts for machinability testing have been attempted, but none satisfy commercial requirements of the material-removal field.

Tool Dynamometer Testing

One of the most widely used procedures for measuring machinability involves the use of a lathe tool dynamometer. This unit, in its simplest form, measures the cutting

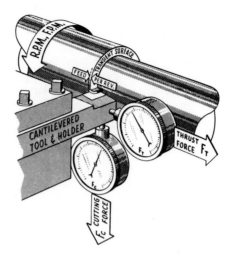

Fig. 6–2.—Force measurement methods for a mechanical lathe dynamometer.

and thrust forces acting on the tool as shown in Fig. 6–2. Three component dynamometers are also available which measure an additional force—the end thrust on the tool. The toolholder can be made very rigid yet sensitive by using strain gages or transducers. Dynamometers are also employed in drilling to measure the thrust and torque on the drill. A schematic drawing of a drill dynamometer is shown in Fig. 6–3.

Having measured the forces, it is then possible to calculate many of the factors involved in the mechanics of chip formation. These factors include the following:

1) Coefficient of friction between the chip and the tool
2) Shearing strain undergone by the metal
3) Work done in overcoming friction between the chip and the tool per unit volume of metal removed
4) Average shear angle of the narrow region in which the process of internal shear occurs during chip formation
5) Other factors important in the metal-cutting process.

However, since tool life in terms of tool wear is not included as one of the above measurements, the data obtained are not adequate for evaluating the machinability of the metal being cut or for determining the machining conditions that should be used in a particular machining operation. Nevertheless, attempts have been made to develop this type of information from tool-force measurements.

On the other hand, tool-force measurements are extremely important in determining power requirements and machining accuracy. Horsepower requirements can be calculated directly from the tangential or cutting force, while the accuracy of the machined component is primarily affected by the normal force—the radial force between the tool and the workpiece. Investigations into the relationship of tool wear to the force components have also been made in an effort to use dynomometer force measurements to control tool wear as part of an adaptive control system.

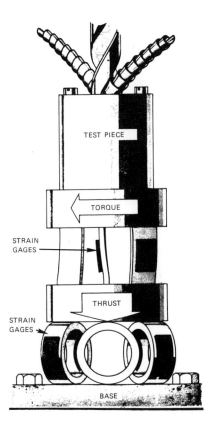

Fig. 6–3.—Force measurement methods for a drill dynamometer.

Thermocouple Testing

The magnitude of the temperature at the cutting edge of the tool has also been used by some as a measurement of the machinability of the material being cut. In this method, a thermocouple is formed by the contact area between the work and the tool, and a voltage proportional to the contact temperature at the cutting tool is generated by this thermocouple. The temperature developed at the cutting edge of the tool is extremely important, because the tool life varies as an exponential function of the cutting temperature. As the temperature increases, the ability of the tool to resist abrasive wear decreases. It is not unusual for the hardness of an HSS tool to change from 64 R_c at room temperature to 57 R_c during cutting as a result of the high temperature developed at the cutting edge. Also, as the tool wears and the rubbing of the cutting edge against the workpiece increases, the temperature climbs until it eventually reaches a point at which an HSS tool will become overtempered and will no longer cut metal. Hence, the temperature developed at the cutting edge does have a very definite effect on the rate of wear of the tool.

There is also another very important factor that influences tool wear—the abrasive nature of the microconstituents in the metal being cut. Since this factor is not taken into account either in the measurements of the tool-tip temperatures or in the resulting tool wear, the method is not satisfactory for determining the machinability of a metal, nor should it be used for determining machining conditions.

Accelerated Tool-Life Testing

Another type of test that has been widely used over the years has been the accelerated tool-life test employing Taylor's equation, $VT^n = C$. As discussed in earlier chapters, this equation produces a straight line when plotted on log-log graph paper. In order to find longer tool-life values, such as 30 min with carbide and 60 min with HSS tools, without actually running tests, the straight line plot of Taylor's equation is used. The procedure is to obtain actual tool-life values of approximately 2 min and 10 min and then to plot the two test points on a log-log chart. The straight line through the two points is then extrapolated into the region where practical tool-life values can be obtained, the assumption being that the curve is a straight line throughout the entire range of cutting speeds.

As far back as 1945, however, it was found by Dr. Michael Field that a break in the curve of the practical cutting-speed range usually occurs when cast iron is machined with carbide tools. As a matter of fact, it was found that tool life on the cast iron used actually decreased when the cutting speed was reduced below 300 sfpm, as shown in Fig. 6–4. Other researchers have also encountered considerable variations in the straight-line theory for steels. Drilling and tapping of many steel alloys, if the cutting speed is too low, results in poor chip formation and low tool life. Also, the reliability of short tool-life tests is generally not very good; the slope of the straight line is question-

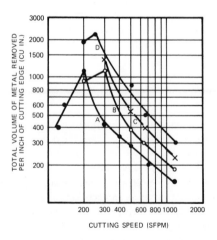

Fig. 6–4.—Tool-life curves for face milling cast iron with a carbide cutter. Curve A, medium pearlite with 5 percent free carbides at 240 BHN; curve B, fine pearlite at 217 BHN; curve C, medium pearlite at 207 BHN; curve D, coarse pearlite at 195 BHN.

able at best, and the results obtained by means of the application of Taylor's equation on a log-log chart can be misleading.

Constant-Pressure Testing

One method that has been used to obtain machining data quickly is the use of a constant-pressure lathe or a penetration-rate measurement in drilling. In both of these procedures, the answer obtained is based on the rate at which the tool moves through a workpiece when a constant force is applied to it. In effect, the procedure involves measurement of the feed rate that is obtained for a given cutting force, the logic for this being that the easier it is to penetrate the workpiece, the more machinable that workpiece must be. Again, since tool life is not involved in this test, the method cannot be used reliably for predicting tool-life values on a variety of alloys. It has, however, been used successfully in screening tests on a given alloy for which the variable was the free-machining additive.

TOOL-LIFE TESTS

The tool-life test is a procedure which involves a minimum amount of opinion in developing speeds, feeds, and other machining parameters under the conditions to be used in production operations. This type of test provides the most reliable data available. In setting up a series of tool-life tests, it is essential that only one variable—cutting speed, feed, tool material, tool geometry, or cutting fluid—be altered at a time. It is also important that the tool or work materials be from a single heat and have uniform microstructures and hardnesses if they are not to be the variables. Variations in tool or work materials can lead to erratic results.

One of the most common tool-life tests uses a single feed rate to develop the relationship between tool life and cutting speed. Once this relationship is determined,

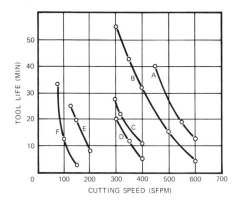

Fig. 6–5.—Tool-life curves for turning AISI 4340 steel with carbide tools. Curve A, spheroidized at 206 BHN; curve B, annealed at 221 BHN; curve C, normalized at 321 BHN; curve D, quenched and tempered at 400 BHN; curve E, quenched and tempered at 500 BHN; curve F, quenched and tempered at 515 BHN.

another factor is then changed and an additional tool-life curve is obtained at the same feed (assuming that feed is not the variable). A family of such tool-life curves, presented in Fig. 6–5, was obtained in this manner for an AISI 4340 steel in six different heat-treated conditions. Considerable machining information can be obtained from this series of curves. Note, for example, that for a tool life of 30 min, the steel in the annealed condition (curve B, 221 BHN) can be machined more than 30 percent faster than the normalized steel (curve C, 321 BHN). Also, the cutting speeds for a tool life of 30 min in turning this steel can range from 80 sfpm (quenched and tempered to 515 BHN) to 490 sfpm (spheriodized) depending on the heat-treated condition. It is obvious from this test that a single machinability index cannot be applied correctly to a single grade of steel. For a similar set of tool-life curves, shown in Fig. 6–6, the test variable was cutting fluid.

Common practice has been to obtain tool-life values at three different cutting speeds for a tool-life curve. An attempt is usually made to select cutting speeds which will provide tool-life values of 5 min, 15 min, and 30 min with carbide tools and 10 min, 30 min, and 60 min with HSS tools. If feed is the test variable, then a curve of the relationship between tool life and feed may be determined for a given cutting speed. A curve of this type is presented in Fig. 6–7 for face milling a thermal-resistant alloy.

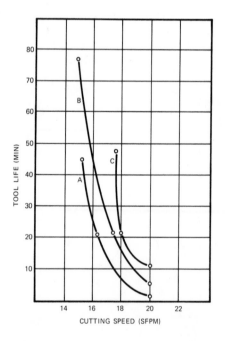

Fig. 6–6.—Tool-life curves for turning thermal-resistant Udimet 500 at 360 BHN using three cutting fluids with an HSS tool. Curve A, highly chlorinated oil; curve B, highly sulfurized oil; curve C, soluble oil, 20:1. (Udimet is a registered trademark of Special Metals Corporation).

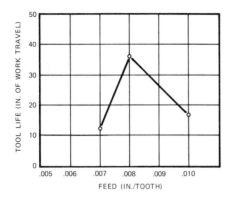

Fig. 6–7.—Effect of feed on tool life in face milling a thermal-resistant alloy at 42 R$_c$.

Bar charts are also used when only one tool-life test point is obtained for each set of conditions. Note, for example, in Fig. 6–8 that the bar chart shows the performances of various types of taps for tapping a titanium alloy.

Tool Wear

The validity of the test results depends to a great extent on the use of a common criterion for the tool-life end point throughout each series of tests. It is common practice in the United States to use the wear land on the flank of the tool as this end point. The sketch in Fig. 6–9 illustrates the location of the wear land and how it is measured on the lip of the tool. In tapping, a go–no go gage is used to determine the tool-life end point.

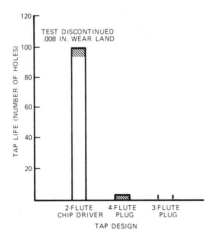

Fig. 6–8.—Results of tapping titanium alloy at 400 BHN with three different taps.

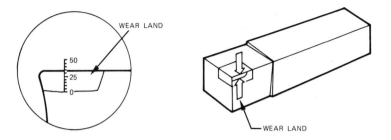

Fig. 6-9.—Wear land measurement in tool-life testing—left, appearance of the wear land under a microscope; right, life size.

Generally, in the shop, the tools used in turning should be changed when the wear land has developed to a width of .030 in. on carbide tools and .060 in. on HSS tools or when the tool fails, whichever occurs first. However, when machining data is being developed by actual tool-life tests with carbide tools it is common practice to use a wear land of .015 in. as the stopping point. This procedure minimizes the amount of material and time required to run the test. This is particularly important when carbide tools are used, because the machining rates for these types of tools are usually high. The data developed by these tests are somewhat conservative and thus should accommodate the wide range of machining conditions that are often found in the shop. The tool-life end point for HSS tools is .060 in. in the tests. Actually, the wear land used in testing varies from one operation to another as shown in Table VI–4.

The machining recommendations listed in Table VI–5 were derived from the tool-life tests just described. Note that not only are the speeds and feeds given, but also the tool geometry, tool material, and even the tool-life results. The data provided by tool-life tests permit the user to proceed with confidence; he does not have to be conservative in selecting machining conditions.

Starting Data

There is no substitute for machining data generated as a result of tool-life testing. Unfortunately, the number of tool-life test data which have been made available represent only a small proportion of the total requirements for all the materials which are

Table VI-4. Wear Land Widths for Testing Various Operations.

Operation	Wear Land Width (in.)	
	HSS	Carbide
Turning	.060	.015 uniform .030 localized
End Milling (Peripheral or Slotting)	.012 uniform .020 localized	.012 uniform .020 localized
Face Milling	.060	.015 uniform .030 localized
Drilling	.015	.015
Reaming	.006	.006
Tapping	Undersize thread to a Class 3 go–no go gage	

1.10 END MILLING

PROFILING

MATERIAL	HARD-NESS BHN	CONDITION	DEPTH OF CUT in.	SPEED fpm	FEED - Inches Per Tooth CUTTER DIAMETER - Inches				TOOL MATL.	SPEED fpm	FEED - Inches Per Tooth CUTTER DIAMETER - Inches				TOOL MATL.
					1/4	1/2	3/4	1 to 2			1/4	1/2	3/4	1 to 2	
STAINLESS STEELS (cont.)	375	Q&T	.050	35	.0005	.001	.002	.003	T15*	90	.001	.002	.003	.004	C-2
	to 425		.015	45	.0005	.0007	.001	.002	T15*	120	.001	.001	.002	.003	C-2
Martensitic 418 440B 440A 440C	48R$_c$	Q&T	.050	20	-	.0005	.0005	.0015	T15*	60	-	.0005	.002	.002	C-2
	to 52R$_c$		.015	25	.0005	.0005	.0005	.001	T15*	80	.0005	.0005	.001	.001	C-2
	54R$_c$	Q&T	.050	-	-	-	-	-	-	35	-	.0005	.001	.0015	C-2
	to 56R$_c$		.015	-	-	-	-	-	-	45	.0005	.0005	.0005	.001	C-2
PRECIPITATION HARDENING STAINLESS STEELS	150	Ann.	.050	70	.001	.002	.003	.004	M2 M7	275	.001	.002	.004	.006	C-2
	to 200		.015	90	.0005	.001	.002	.003	M2 M7	360	.0005	.001	.003	.005	C-2
	275	Hard.	.050	55	.001	.002	.003	.004	M2 M7	225	.001	.002	.004	.006	C-2
	to 325		.015	70	.0005	.001	.002	.003	M2 M7	290	.0005	.001	.003	.005	C-2
17-4PH 17-7PH PH15-7Mo AM350 AM355 AM359	325	Hard.	.050	50	.0005	.001	.002	.003	M2 M7	200	.001	.002	.003	.004	C-2
	to 375		.015	65	.0005	.0007	.001	.002	M2 M7	260	.001	.001	.002	.003	C-2
	375	Hard.	.050	40	.0005	.001	.002	.003	M2 M7	75	.001	.002	.003	.004	C-2
	to 440		.015	50	.0005	.0007	.001	.002	M2 M7	100	.001	.001	.002	.003	C-2
TITANIUM ALLOYS Commercially Pure 99.5	110	Ann.	.050	125	.002	.004	.006	.007	M2 M7	325	.002	.003	.004	.006	C-2
	to 170		.015	150	.0015	.004	.006	.007	M2 M7	375	.001	.003	.004	.006	C-2
Commercially Pure 99.2 99.0 0.15 to 0.20 Pd	140	Ann.	.050	120	.002	.004	.006	.007	M2 M7	300	.002	.003	.004	.006	C-2
	to 200		.015	150	.0015	.004	.006	.007	M2 M7	375	.001	.003	.004	.006	C-2
Commercially Pure 99.0 98.9	200	Ann.	.050	60	.0015	.003	.005	.006	M2 M7	160	.0015	.003	.006	.008	C-2
	to 275		.015	75	.001	.002	.004	.005	M2 M7	190	.001	.002	.005	.007	C-2
Alpha & Alpha-Beta Alloys 2Fe-2Cr-2Mo 5Al-2.5Sn 5Al-2.5Sn (low O) 7Al-2Cb-1Ta 4Al-3Mo-1V	300	Ann.	.050	55	.0015	.003	.005	.006	M2 M7	140	.0015	.003	.006	.008	C-2
	to 340		.015	70	.001	.002	.004	.005	M2 M7	175	.001	.002	.005	.007	C-2

Fig. 6–10.—Example of profile end-milling data from the *Machining Data Handbook* (5).

Table VI-5. Recommended Cutting Conditions for Machining René 41* Solution-Treated to 321 BHN.

Operation	Tool Material	Tool Geometry	Tool Used for Tests	Depth of Cut (in.)	Width of Cut (in.)	Feed	Cutting Speed (sfpm)	Tool Life	Wear Land (in.)	Cutting Fluid
Turning	C-2 Carbide	BR: 0° SCEA: 15° SR: 5° ECEA: 15° Relief: 5° NR: 1/32 in.	1/2-in.-square throwaway holder with mech. chip breaker	.062	——	.009 ipr	70	40 min	.016	soluble oil (1:20)
	T-15 HSS	BR: 0° SCEA: 0° SR: 15° ECEA: 5° Relief: 5° NR: 1/32 in.	5/8-in.-square tool bit	.062	——	.009 ipr	12	75 min	.010	soluble oil (1:20)
Face Milling	C-2 Carbide	AR: 0° TR: 5° RR: 7° Incl: -5° CA: 45° ECEA: 5° Clearance: 10°	4-in.-diameter face mill	.060	2	.0065 in./tooth	50	28 in./tooth work travel	.030	highly chlorinated oil
	T-15 HSS	AR: 0° TR: 14° RR: 20° Incl: -14° CA: 45° ECEA: 5° Clearance: 10°	4-in.-diameter face mill	.060	2	.011 in./tooth	22	80 in./tooth work travel	.012	highly chlorinated oil
End-Mill Slotting	M-2 HSS	30° RH helix RR: 10° CA: 45° Peripheral cl.: 10° ECEA: 3°	3/4-in.-dia. 4-tooth end mill 1 in. flute length	.250	3/4	.002 in./tooth	22	100 inches work travel	.020	soluble oil (1:20)

Table VI-5. (*Continued*)

Slot Milling Down Milling	C-2 Carbide	AR: −5° bi-neg. RR: 5° ECEA: 1° CA: 45° × .030 in. Clearance: 10°	.125	1	.003 in./tooth	94	48 in./tooth work travel	.030	highly chlorinated oil
Drilling	T-15 HSS	1/4-in.-dia. heavy web type drill, 2-1/2-in. O.L., 1-1/2-in. flute length	1/2-in. thru hole	—	.002 ipr	17	90 holes	.020	highly chlorinated oil
Tapping	M-10 HSS	2-flute plug tap, spiral pt., 75 percent threaded	5/16-24 NF plug tap	—	—	12	98 holes	Tap breakage	highly chlorinated oil
Reaming	M-2 HSS	6-flute straight chucking reamer CA: 45° Clearance: 10°	.272-in.-diameter reamer	1/2-in. thru hole	.005 ipr	25	95 holes	.016	highly chlorinated oil

*[TM] Vacuum Melted Alloys, Metallurgical Products Department, General Electric Company. Nominal chemical composition (percent): Cr, 19.0; Co, 11.0; Mo, 10.0; Fe, 5.0; Ti, 3.0; C, .10; Al, 1.5; Ni; balance.

being machined today. When tool-life data cannot be supplied, starting recommendations are often available as a result of limited testing or experience. Most recently, a systematic collection of starting data has been compiled through effort initiated by the United States Army Rock Island Arsenal Laboratory. Their program has resulted in the *Machining Data Handbook* (5), which contains machining data for all important commercial alloys and for virtually all machining operations. A typical format for turning data from this handbook is shown in Fig. 2–10. Other typical formats for end milling and drilling data are shown in Figs. 6–10 and 6–11. Data for the handbook were collected with the help of the United States Army Weapons Command Metal Cutting Advisory Committee and approximately 250 companies in industry, and, as such, the data gathered reflect the state of machinability data collection methods as they exist today.

While these data portray our present knowledge rather faithfully, it should by no means be concluded that they are optimum data. On the contrary, they are purposely specified at less than optimum values in order to make them generally applicable throughout industry, considering variations in capability, sophistication of machine-tool equipment, and other plant conditions. Data for the handbook, however, have been carefully screened to ensure that they provide good nominal starting conditions for both N/C and conventional manufacturing methods.

From Figs. 6–10 and 6–11 it is evident that specific machining recommendations are listed. These are chosen corresponding to a tool life of approximately 1 to 2 hours of cutting time for most of the common alloys when HSS or brazed carbide tools are used. A tool life of 30 min to 60 min is used for disposable carbide tools. In actual shop operations, tool life might be three to four times these values because of the loading and unloading times associated with most machining operations.

Many shops throughout the country have low production requirements but machine a wide variety of materials. For this type of manufacturer, a collection of reliable starting recommendations may fulfill most of his data requirements and could well be optimum for his needs.

Compromises were made in providing starting recommendations in the *Machining Data Handbook* as a result of the groupings of similar alloys. For example, data have not as yet been generated to justify different starting recommendations for alloy steels such as 1320, 4320, 5120, and 8720; and stainless steels 403, 410, 420, and 501 are grouped under one material heading. However, in each of these cases, machining data vary depending upon the hardness of the alloy being machined. There are variations within material groups such as those noted above, but insufficient data exist to justify selection of different starting recommendations.

MACHINABILITY DATA FOR NUMERICAL CONTROL

While the operation of many shops does not justify sophisticated data, there are many other production facilities in which starting recommendations no longer can be used to meet competition. In many shops, N/C equipment is being used to an increasing extent, while in others cost and production considerations for conventional equipment are so critical that every effort must be made to seek optimized machining conditions. Machining data of this type are classified as "machinability data for numerical control." As has been pointed out before in this book, these data are not necessarily limited to application to N/C machines; rather, the designation indicates the absolute necessity of having this caliber of data when machining is monitored by numerical

1.16 DRILLING

MATERIAL	HARD-NESS BHN	CONDITION	SPEED fpm	FEED - Inches Per Revolution								HSS TOOL MATERIAL
				NOMINAL HOLE DIAMETER - Inches								
				1/16	1/8	1/4	1/2	3/4	1	1-1/2	2	
ALLOY STEELS (cont.)	45R$_C$ to 48R$_C$	Quenched and Tempered	20	-	.0005	.001	.002	.002	.003	.003	.004	T15 M33
1330 4337 6260 1332 4340 6270 1335 4640 6290 1340 50B40 6342 . . . (SEE PAGE 224 FOR COMPLETE MATERIAL LIST)	48R$_C$ to 50R$_C$	Quenched and Tempered	20	-	.0005	.001	.002	.002	.003	.003	.004	T15 M33
	50R$_C$ to 52R$_C$	Quenched and Tempered	15	-	.0005	.001	.002	.002	.003	.003	.004	T15 M33
NITRIDING STEELS Nitralloy 125H Nitralloy 135G Nitralloy 135M Nitralloy N Nitralloy 230	200 to 250	Annealed	50	-	.002	.003	.005	.008	.010	.011	.013	M10 M7 M1
	300 to 350	Normalized or Quenched and Tempered	35	-	.002	.003	.006	.008	.009	.010	.011	T15 M33
ARMOR PLATE MIL-A-1260 (ORD)	250 to 320	Quenched and Tempered	25	-	.001	.002	.003	.005	.006	.007	.008	T15 M33
ULTRA-HIGH STRENGTH STEELS	200 to 250	Annealed	55	-	.002	.003	.005	.008	.010	.011	.013	M10 M7 M1
	250 to 300	Normalized	50	-	.002	.003	.006	.008	.009	.010	.011	M10 M7 M1
D6AC MX-2 4340	43R$_C$ to 48R$_C$	Quenched and Tempered	20	-	.001	.002	.003	.003	.004	.005	.005	T15 M33
	48R$_C$ to 50R$_C$	Quenched and Tempered	20	-	.0005	.001	.002	.002	.003	.003	.004	T15 M33
	50R$_C$ to 52R$_C$	Quenched and Tempered	15	-	.0005	.001	.002	.002	.003	.003	.004	T15 M33

Fig. 6–11.—Example of drilling data from the *Machining Data Handbook* (5).

control. It is quite evident that large capital investments in this type of equipment make optimum metal removal more essential.

A mitigating circumstance in N/C operations is the requirement to employ programmers with limited machining experience. An experienced manufacturing engineer may have a fund of experience upon which N/C programming might be based. As such, he may not find it essential to rely on handbook data, at least in the machining of common materials such as ordinary steels, cast irons, or aluminum alloys. However, with the extensive growth of N/C equipment at the present time, there are relatively few manufacturing engineers with programming experience. As a result, personnel with little experience in machining but with training in data processing procedures do the programming. These men must rely heavily upon machining data, and therefore reliable data must be made available to them.

Machinability data for numerical control are based upon the data collected from tool-life tests. One source of data slanted towards N/C methods is available from the Air Force Machinability Data Center (AFMDC) in Cincinnati, Ohio, as a result of programs sponsored by the Air Force Manufacturing Technology Division (AFMTD) at Wright-Patterson Air Force Base, Ohio. Operations studied by the AFMTD are listed in a report entitled *Machining Data for Numerical Control* (designated AFMDC 66-1 and its supplement, 68-2) and include machining data for N/C in turning, face milling, drilling, peripheral end milling, end-mill slotting, tapping, and reaming (6, 7). Materials described include the more important alloy steels, ultrahigh-strength steels, tool steels, stainless steels, nickel- and cobalt-based high-temperature alloys, refractory alloys, and a few nonmetallic materials. Typical data from *Machining Data for Numerical Control* are shown in Figs. 4–1 through 4–3, 4–13, and 4–16 in Chapter 4.

The computer approach to extensive cost and production-rate optimization discussed in Chapter 4 is based upon machining data obtained from *Machining Data for Numerical Control*. These data are of considerable assistance because they express the relationship of tool life to cutting speed. They are of great value to any type of machining situation of the materials for which data are presented. However, even for applications involving N/C tools, starting recommendations or other information based on tests or experience must be used to cover other material types and the more than thirty other machining operations for which systematic tool-life data have not been developed. Nevertheless, machinability data of the type described here are the type most useful to engineers, designers, planners, parts programmers, and production specialists.

Analyses which have been made of the high costs of material removal and the cost savings resulting from intelligent application of data justify the manufacturer's interest in developing the data necessary for optimization. In many sectors of industry, the trend is toward increasing use of hard-to-machine materials, as shown in Table VI–6. Fig. 6–12 shows the trends in machining costs that result from use of the more difficult-to-machine materials (8).

Cost reductions resulting from applications of machining data are easy to develop for individual situations within a manufacturing facility. However, in order to provide management with a general incentive to institute programs for the application of data, it becomes necessary to resort to simple methods for estimating the order of magnitude of machining costs and of potential cost savings. The calculations shown in Table VI–2 were necessarily based upon overall estimates; however, within a given manufacturing facility, management does have reliable data on factors such as the number of machine tools, the number of hours in an average working day, the number of working days,

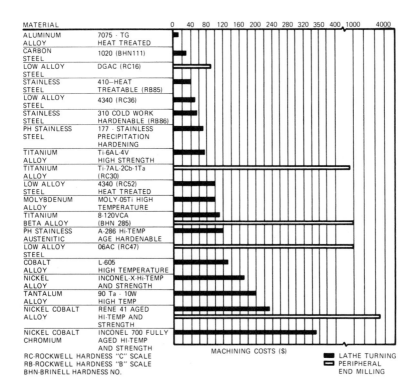

Fig. 6-12.—Machining costs for hard-to-machine materials. Chart is based on machining cost for a part made from 7075–T6 aluminum bar stock (machinability—100 percent) and a 60-min tool life. Approximate costs do not reflect the costs of materials, cutting tools, or heat treatment (8).

and labor costs. Total annual costs for material removal calculated in this way are generally large enough to excite management's attention, leading to increased demand for detailed machining data based on tool-life studies.

Tool life and other machinability data have far-reaching significance. They provide information from which production rates can be calculated, and these rates, in turn, affect the number of machine tools required, floor space, individual plant facilities, and

Table VI-6. Survey of Hard-to-Machine Material Usage Percentages, 1964–1974 (8).

Material	1964 Percentage	1969 Percentage	Estimated 1974 Percentage
High-tensile steels	15.2	26.2	22.0
Titanium to 250 ksi	8.8	16.2	23.0
Refractories to 160 ksi	6.9	5.2	9.0
Filament to 1,000 ksi	1.4	4.0	8.0
Composite to 400 ksi	3.0	2.6	7.2
Total	35.3	54.2	69.2

labor requirements. In a modern approach to material removal, machining data are required by manufacturing engineering supervisors, tool and process engineers, tool designers, industrial engineers, time-study engineers, production planners, machine-shop supervisors, foremen and machine-tool operators, value analysis engineers, N/C programmers and planners, and purchasing agents. Purchasing agents can no longer intelligently purchase machine tools, cutting tools, cutting fluids, or even work materials without having adequate machining data. Currently, some aggressive purchasing agents, realizing the vast expenditures for cutting tools, have initiated testing to develop data for vendor control. Advanced manufacturing technology will be required to make use of machinability data more and more in order to minimize the cost of the machining operation itself as well as to reduce the almost total material loss which results from chip making. Machinability data, in the broadest sense, will contribute extensively toward minimizing the amount of material removal which is actually necessary in manufacturing. In other words, machining data will contribute toward development of processes which will eliminate much material waste.

REFERENCES

1. Hans Ernst, "Economic Importance of Chip Making," *ASTME Paper No. MM58-101* (1958).
2. *The Tenth Annual American Machinist Inventory of Metalworking Equipment 1968*, Part 2, Nov. 18, 1968, New York, McGraw-Hill, Inc., p. 4.
3. National Machine Tool Builders Association, *Economic Handbook 1970/71 of the Machine Tool Industry*, Washington, D.C., National Machine Tool Builders' Show, Inc., 1970.
4. United States, Department of Commerce, Business and Defense Services Administration, *The Metal-Cutting Tool Industry 1965–1969*, Washington, D.C., U.S. Government Printing Office, Aug., 1970.
5. Metcut Research Associates, Inc., *Machining Data Handbook*, 2d ed., Cincinnati, 1971.
6. Michael Field, Clarence Mehl, and John F. Kahles, *Machining Data for Numerical Control (AFMDC 66-1)*, Cincinnati, Air Force Machinability Data Center, 1966.
7. Michael Field, Clarence Mehl, and John F. Kahles, *1968 Supplement to Machining Data for Numerical Control (AFMDC 68-2)*, Cincinnati, Air Force Machinability Data Center, 1968.
8. United States, Department of the Air Force, "Profile Milling Requirements for the Hard Metals," report of the Ad Hoc Machine Tool Advisory Committee to the Department of the Air Force, 1965, pp. 25, 28.

APPENDIX

AIR FORCE
MACHINABILITY
DATA CENTER

The Air Force Machinability Data Center (AFMDC), Cincinnati, Ohio, is operated for the Air Force Materials Laboratory, Materials Support Division, by Metcut Research Associates, Inc., under contract F33615-71-1112.

AFMDC collects, evaluates, stores, and disseminates specific and detailed machinability data for the benefit of industry and government. AFMDC emphasizes engineering evaluation of its information in order to develop optimized material-removal parameters—speeds, feeds, depths of cut, tool material and geometry, cutting fluids, and other significant variables. Data are processed by AFMDC for all types of materials and for material-removal operations such as turning, milling, drilling, tapping, grinding, electrical discharge machining, and electrochemical machining.

AFMDC's data file contains over 27,000 selected documents pertaining to all phases of material-removal technology. This data file is supported by an automated retrieval system controlled by an IBM 1130 computer. Information is retrieved based upon the specific material (with definite chemical, physical, and mechanical properties) and the specific material-removal operation to be used. Sophisticated computerized search techniques are employed which use a combination of search parameters to produce source data from the file. The search parameters entered into the computer are controlled to satisfy the requirements of any specific inquiry.

Intensive effort is maintained by AFMDC in order to serve as a communication link for government and industry by providing material-removal-related services. Major emphasis is placed on providing analyzed data in answer to technical inquiries. Services are available to the aerospace industry; all Department of Defense agencies and their contractors; and other government agencies, technical institutions, and nonmilitary industries in a position to assist the defense effort. At present, inquiry service is available without charge; however, AFMDC is conducting studies into possible future collection of output costs related to this service. AFMDC also maintains a selected mailing list of users for the dissemination of new information or services available from the Center.

To assist AFMDC in retrieving data in answer to an inquiry, the user should provide the following information:

1) The material to be machined (specification or trade name), its condition (cast,

hot rolled, cold drawn, annealed, quenched and tempered, etc.), its microstructure, and its hardness.

2) The material-removal operation (turning, milling, drilling, tapping, surface grinding, electrical discharge machining, electrochemical machining, etc.) to be performed.

3) Specific reasons the data is required (unless needs are proprietary). This information allows AFMDC to broaden the scope of its technical advice.

4) Specific delivery requirements.

5) The person or office to whom the reply should be sent.

6) All details concerning present practices, including feeds, speeds, cutting-tool material and geometry, cutting fluids, etc., in the event the inquiry pertains to improvement of an existing machining situation.

To request specific machining information . . .

Telephone: 513-271-9510

TWX: 810-461-2840

Write: Air Force Machinability Data Center
3980 Rosslyn Drive
Cincinnati, Ohio 45209

Association of the names of companies and individuals with specific requests is kept confidential by AFMDC. However, data developed remain the property of AFMDC for dissemination as required to answer other inquiries and to develop data publications.

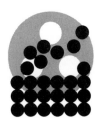

BIBLIOGRAPHY

Ameday, A. A. "Management Systems for Machine Tool Utilization," *ASTME Paper No. MM68-702* (1968).

American Society of Tool and Manufacturing Engineers. *Machining with Carbides and Oxides.* New York: McGraw-Hill Book Company, Inc., 1962.

———. *Manufacturing and Estimating Handbook.* New York: McGraw-Hill Book Company, Inc., 1963.

———. *Numerical Control in Manufacturing.* New York: McGraw-Hill Book Company, Inc., 1963.

———. *Tool Engineers Handbook,* 2d ed. New York: McGraw-Hill Book Company, Inc., 1959.

Anderson, P. J., and P. F. Boyer. "Effective Utilization of Numerically Controlled Turning Machines at the Oak Ridge Y-12 Plant," *SME Paper No. MR70-199* (1970).

A Treatise on Milling and Milling Machines, 3rd ed. Cincinnati: Cincinnati Milling Machine Company, 1951.

Berra, P. Bruce, and Moshe M. Barash. "Automated Planning and Optimization of the Production Process in the Manufacture of Discrete Metal Products," *Report No. 6,* Purdue Laboratory of Applied Industrial Control, Purdue University, Lafayette, Indiana.

———. "The Automated Planning and Optimization of Manufacturing Processes." Paper presented at the International Optimal Systems Planning Symposium, Cleveland, June 20-22, 1968.

Bhattacharyya, Amitabha, and Inyong Ham. *Design of Cutting Tools: Use of Metal Cutting Theory.* Dearborn, Mich.: American Society of Tool and Manufacturing Engineers, 1969.

Black, P. H. *Theory of Metal Cutting.* New York: McGraw-Hill Book Company, Inc., 1961.

Brewer, R. C. "On the Economics of Basic Turning Operation," *ASTME Paper No. 57-A-58,* February, 1958.

Brierley, Robert G. "Cemented Oxides: Where, When and How to Apply," *SAE Paper No. 981 B,* January, 1965.

———, and H. J. Siekmann. *Machining Principles and Cost Control.* New York: McGraw-Hill Book Company, Inc., 1964.

Brown, R. H. "Oh the Selection of Economical Machining Rates," *The International Journal of Production Research,* Vol. 1, No. 2 (March, 1962).

Coated Abrasives Manufacturers' Institute. *Coated Abrasives: Modern Tool of Industry,* 1st ed. New York: McGraw-Hill Book Company, Inc., 1958.

Colding, Bertil N. "Machinability of Metals and Machining Costs," *International Journal of Tool Design and Research,* Vol. 1 (1961).

———. "Machining Economics and Industrial Data Manuals." Paper presented at C.I.R.P. Annual Meeting, University of Manchester, Institute of Science and Technology, Manchester, England, September, 1968.

Conn, H. "Tomorrow's Machining Concept—Optimization " *Automatic Machining,* May, 1957.

Cook, Nathan H. *Manufacturing Analysis.* Reading, Mass.: Addison-Wesley Publishing Company, Inc., 1966.

Corwin, Daniel W. "A Computerized Approach to the Least-Cost Optimization of Job Shop Production," B.S. thesis in mechanical engineering, Massachusetts Institute of Technology, May, 1967.

Datsko, Joseph, and O. W. Boston. "Relative Abrasiveness of the Cast Surfaces of Various Gray-Iron Castings on Single-Point Tools of High-Speed Steel," *Transactions of the ASME,* Vol. 75 (1953).

Downing, C. E. "Setting Machining Feeds and Speeds: Optimization and Machining Economics," *ASTME Paper No. MR62-152* (1962).

Doyle, L. E. *Manufacturing Processes and Materials for Engineers.* Englewood Cliffs, N.J.: Prentice-Hall, Inc., 1961.

DuSault, Roland A. "Industrial Engineering in a Systems Manufacturing Plant," *ASTME Paper No. MM67-682* (1967).

Engelskirchen, W. H. "Anpassung von Programmiersprachen der Fertigungstechnik an numerisch gesteuerte Werkzeugmaschinen (Adaptation of Production-Oriented Programming Languages to N/C Machine Tools)." Doctoral thesis, University of Aachen, Germany, 1968.

Ermer, D. G., and R. Faria-Gonzalez. "An Analytical Sensitivity Study of the Optimum Machining Conditions," *ASME Paper No. 67-WA/PROD-20* (1967).

Ernst, Hans. "Economic Importance of Chip Making," *ASTME Paper No. MM58-101* (1958).

————. "Economics of Machining," *Cincinnati Milling,* Vol. 13, No. 1. The Cincinnati Milling Machine Company, 1956.

————, and Michael Field. "Speed and Feed Selection in Carbide Milling with Respect to Production, Cost, and Accuracy," *Transactions of ASME,* Vol. 68, April (1946).

Feldman, C. G. "Automatic Data Processing for Numerically Controlled Machine Tools." Paper presented at the ASME Production Engineering Conference, Milwaukee, Wisconsin, May 18, 1960.

Field, Michael, and A. F. Ackenhausen. *Determination and Analysis of Machining Costs and Production Rates Using Computer Techniques, (AFMDC 68-1).* Cincinnati: Air Force Machinability Data Center, 1968.

————, and John F. Kahles. *Grinding Ratios for Aerospace Alloys, (AFMDC 66-2).* Cincinnati: Air Force Machinability Data Center, 1966.

————, Clarence Mehl and John F. Kahles. *Machining Data for Numerical Control (AFMDC 66-1).* Cincinnati: Air Force Machinability Data Center, 1966.

————, ———— and ————. *Supplement to Machining Data for Numerical Control (AFMDC 68-2).* Cincinnati: Air Force Machinability Data Center, 1968.

————, Norman Zlatin, Roy L. Williams, and Max Kronenberg. "Computerized Determination and Analysis of Cost and Production Rates for Machining Operations: Part I—Turning," *Transactions of the ASME,* Vol. 90, August (1968).

————, et al. "Computerized Determination and Analysis of Cost and Production Rates for Machining Operations," *ASME Paper No. 67-WA/PROD-18* (1967).

————, et al. *Machining of High Strength Steels with Emphasis on Surface Integrity (AFMDC 70-1).* Cincinnati: Air Force Machinability Data Center, 1970.

Final Report on Machinability of Materials (AFML-TR-444), Vol. 6. Cincinnati: Metcut Research Associates, Inc., 1966.

Fischel, Terry J. "N/C Machine Utilization Monitoring System," *SME Paper No. MR70-200* (1970).

Gardiner, K. M. "Computer Decides Conditions for Minimum Cost Machining," *Metalworking Production,* Vol. 109, No. 49 (December 8, 1965).

General Catalog #GT0-240: Carboloy Cemented Carbides and Cemented Oxides. Detroit: General Electric Company, Metallurgical Products Department, 1970.

Gilbert, W. W. "Economics of Machining," in *Machining: Theory and Practice.* Metals Park, Ohio: American Society for Metals, 1950.

————, and E. J. Weller. "Application of a Machinability Computer," *ASTE Annual Collected Papers,* Paper No. 24T26, 1956.

————, and W. C. Truckenmiller. "Nomograph for Determining Tool Life and Power When Turning with Single-Point Tools," *Mechanical Engineering,* Vol. 65, December (1943).

Golish, R. J. *One Hundred Applications of Grinding to the Aerospace Materials.* Tempe, Ariz.: Industrial Arts and Sciences, 1967.

Ham, Inyong. "Economics of Machining: Analyzing Optimum Machining Conditions by Computers," *ASTME Paper No. MR64-534* (1964).

Hirsch, B. "Automatic Programming for N/C Lathes with EXAPT 2," *Machinery and Production Engineering,* August 6, 1969.

————. "Bestimmung optimaler Schnittbedingungen bei der maschinellen Programmierung von NC-Drehmaschinen mit EXAPT 2 (Determination of Optimal Cutting Values for Computer-Assisted Programming of N/C Lathes with EXAPT 2)." *Industrie-Anzeiger,* Vol. 90, No. 24 (1968).

————. "Ein System zur Ermittlung von Zerspanungsvorgabewerten, insbesondere bei rechnergestützte Programmierung numerisch gesteuerter Drehmaschinen (A System for the Determination of Cutting Values for the Computer-Assisted Programming of N/C Lathes)." Doctoral thesis, University of Aachen, Germany, 1968.

————, and Hans Zölzer. *EXAPT 2 Material File.* Aachen, Germany: EXAPT-Verein, 1968.

Holmes, W. G. *Turning: Tools, Methods, Cost.* Detroit: Reed Technical Service, 1958.

Howe, Raymond E., ed. *Introduction to Numerical Control in Manufacturing.* Dearborn, Mich.: American Society of Tool and Manufacturing Engineers, 1969.

————. *Producibility/Machinability of Space-Age and Conventional Materials.* Dearborn, Mich.: American Society of Tool and Manufacturing Engineers, 1968.

IBM Application Program—System 360 AD-APT/AUTOSPOT Numerical Control Processor (360A-CN-09X): Version 2, Part Programmer Manual. White Plains, N.Y.: International Business Machines Corporation, Technical Publications Department, 1967.

International Industrial Diamond Conference Proceedings, 1969. Moorestown, N. J.: Industrial Diamond Association of America, Inc., 1970.

Jones, W. L., and J. R. Morgan. "Use of Decision Diagrams to Examine Tooling Problems," *The International Journal of Production Research,* Vol. 6, No. 2 (March, 1967).

Kane, G. E. "A Regression Method for Developing Cutting Force Formula," *ASTME Paper No. MR66-212* (1966).

————. "Some Fundamental Considerations Relative to Tool Selection," *ASTME Paper No. MR64-167* (1964).

Kobayashi, Akira. *Machining of Plastics.* New York: McGraw-Hill Book Company, Inc., 1967.

Kronenberg, Max. *Machining Science and Application.* New York: Pergamon Press, 1966.

LeGrand, Rupert, ed. *The New American Machinist's Handbook.* New York: McGraw-Hill Book Company, Inc., 1955.

McCullough, E. M. "Economics of Multitool Lathe Operations," *Transactions of the ASME-B,* Vol. 85, No. 4 (November, 1963).

"Machining," in *Metals Handbook,* Vol. 3, 8th ed. Metals Park, Ohio: American Society for Metals, 1967.

Machining Data Handbook, 2d ed. Cincinnati: Metcut Research Associates, Inc., 1971.

Machining Difficult Alloys. Metals Park, Ohio: American Society for Metals, 1962.

Machining the Space-Age Metals. Dearborn, Mich.: American Society of Tool and Manufacturing Engineers, 1965.

Manual on the Cutting of Metals with Single-Point Tools, 2d ed. New York: American Society of Mechanical Engineers, 1952.

Maranchik, John, Jr. *Machining Data for Titanium Alloys (AFMDC 65-1).* Cincinnati: Air Force Machinability Data Center, 1965.

Mathias, R. A. "Adaptive Control of the Milling Process." Paper presented at the IEEE National Machine Tools Industry Conference, Cleveland, Ohio, October, 1967.

Merchant, M. E. "Mechanics of the Metal Cutting Process," *Journal of Applied Physics,* Vol. 16, No. 5 (May, 1945).

Metal-Cutting Bibliography: 1943-1956. Detroit: American Society of Tool and Manufacturing Engineers, 1960.

Metal Cutting Tool Handbook. New York: Metal Cutting Tool Institute, 1965.

National Machine Tool Builders Association. *Economic Handbook 1970/71 of the Machine Tool Industry.* Washington, D.C.: National Machine Tool Builders' Show, Inc., 1970.

N/C Handbook. Detroit: The Bendix Corporation, Industrial Controls Division, 1967.

Numerical Contouring and Positioning Controls (NEC-1020). Waynesboro, Va.: General Electric Company, June, 1968.

"Numerical Control: Tomorrow's Technology Today," *Proceedings of the Fifth Annual Meeting and Technical Conference of the Numerical Control Society, Philadelphia, April 3-5, 1968.* Princeton, N.J.: Numerical Control Society, 1968.

Oberg, Erik, and F. D. Jones. *Machinery's Handbook,* 18th ed. New York: Industrial Press, 1968.

Okushima, K., and K. Hitomi. "A Study of Economical Machining: An Analysis of the Maximum Profit Cutting Speed," *The International Journal of Production Research,* Vol. 3, No. 1 (January, 1964).

1130 Work Measurement Aids: User's Manual H20-0363-0. White Plains, N.Y.: International Business Machines Corporation, Technical Publications Department, 1967.

Operating Manual for the Carboloy Machinability Computer: Manual No. MC-101-B. Detroit: General Electric Company, Metallurgical Products Department, 1957.

"Operation Sheets from a Computer," *American Machinist,* Vol. 109, No. 23 (November 8, 1965).

Opitz, H., and B. Hirsch. "Programmation automatique des machines-outils à commande numérique (Automatic Programming of N/C Machine Tools)." Paper presented at C.I.R.P. European Conference on the Numerical Control of Machine Tools, Paris, 1968.

————, et al. "Das Programmier system EXAPT (The EXAPT Programming System)." *TZ für praktische Metallbearbeitung,* Vol. 61, No. 6 (1967).

————. *Coding Guidelines and Definitions for the Work-Describing Classifications System.* Essen, Germany: Verlag W. Girardet, 1966.

Part Processing Manual No. TGP-61, Milwaukee-Matic Model II. Milwaukee: Kearney and Trecker Corporation.

Porten, Charles. "Automated Planning of Manufacturing Operations," *Automation,* Vol. 13, No. 5 (May, 1966).

"Process Master for NAS 913, Rev. 2, August 15, 1965—Test 4.3.3.8.5, 3-Axis Profiler." Detroit: The Bendix Corporation, 1965.

Programmer's Manual No. 3—Cintimatic 200 Series Control, Publication No. M-2492-2. Cincinnati: Cincinnati Milling Machine Company.

Reckziegel, Dieter. "N.C. Programming System EXAPT," *Proceedings of 1967 Seminar on Horizons in Manufacturing Technology,* University of Michigan, Institute of Science and Technology, Industrial Development Division, April, 1967.

Romanov, K. F., and K. F. Sotnikova. "Effect of Workshop Conditions on Cutting Speeds," *Russian Engineering Journal,* Vol. XLVI, No. 4 (1966).

Schneck, Donald E. "Feasibility of Automated Process Planning," Ph.D. dissertation, Purdue University, Lafayette, Ind., 1966.

Selection and Application of Single-Point Metal-Cutting Tools: Publication GT9-270. Detroit: General Electric Company, Metallurgical Products Department, 1969.

Shaw, Milton C. *Metal Cutting Principles.* Cambridge, Mass.: The M.I.T. Press, 1967.

————, N. H Cook and P. A. Smith. "Putting Machinability Data to Work," *Tool Engineers,* Vol. 35, No. 2 (August, 1955).

Siekmann, H. J. "Now an Easier Way to Find Best Cutting Speed," *American Machinist,* Vol. 102, No. 3 (February 10, 1958).

Smith, Frank E. "Automatic Clamping as It Affects the Utilization of N/C Machines," *SME Paper No. MR70-201* (1970).

Snider, R. E., and J. F. Kahles. *Machining Data for Beryllium Metal (AFMDC 66-3).* Cincinnati: Air Force Machinability Data Center, 1960.

Speeds and Feeds for Better Turning Results, 4th ed. Sidney, Ohio: Monarch Machine Tool Company.

Springborn, R. K., ed. *Cutting and Grinding Fluids: Selection and Application.* Dearborn, Mich.: American Society of Tool and Manufacturing Engineers, 1967.

————. *Non-Traditional Machining Processes.* Dearborn, Mich.: American Society of Tool and Manufacturing Engineers, 1967.

Swinehart, Haldon J., ed. *Cutting Tool Material Selection.* Dearborn, Mich.: American Society of Tool and Manufacturing Engineers, 1968.

————. *Gundrilling, Trepanning and Deep Hole Machining,* 2d ed. Dearborn, Mich.: American Society of Tool and Manufacturing Engineers, 1967.

Taylor, Frederick W. "On the Art of Cutting Metals," *Transactions of the ASME,* Vol. 28, No. 1119 (1907).

The Tenth Annual American Machinist Inventory of Metalworking Equipment, Part 2, Nov., 18, 1968. New York: McGraw-Hill Book Company, Inc.

Thompson, Jack. "How to Improve Machining Center Productivity Through Better Programming," *Machinery,* Vol. 75, No. 5 (January, 1969).

United States, Department of Commerce, Business and Defense Services Administration. *The Metal-Cutting Tool Industry 1965–1969.* Washington, D.C.: U.S. Government Printing Office, August, 1970.

United States Air Force Machinability Report: Increased Production, Reduced Costs Through a Better Understanding of the Machining Process and Control of Materials, Tools and Machines. 4 Vols. Woodridge, N.J.: Curtiss-Wright Corporation, 1950–60. Reprinted in 1 Vol. by Air Force Machinability Data Center, Cincinnati, 1967.

Weill, R., et al. "The Use of Electronic Computers for the Determination of Optimum Machining Conditions." *Proceedings of the 3rd International M.T.D.R. Conference,* University of Birmingham, Birmingham, England, September, 1962.

Weller, E. J., and C. A. Reitz. "Optimizing Machinability Parameters with a Computer," *ASTME Paper No. MS66-179* (1966).

Wilkie Brothers Foundation. *Fundamentals of Band Machining.* Albany, N.Y.: Delmar Publishers, 1964.

————. *Precision Surface Grinding.* Albany, N.Y.: Delmar Publishers, 1964.

Williams, Roy L., and William B. Johnson. "Applicability of Machining Data to Numerical Control," *ASTME Paper No. MR68-711* (1968).

Witthoff, J. "Die Ermittlung der günstigsten Arbeitsbedingungen bei der spanabhebenden Formgebung (Determination of Optimal Machining Conditions for Metalcutting)," *Werkstatt und Betrieb,* Vol. 85, No. 10 (1952).

———. "Ergänzende Betrachtungen zur Ermittlung der güngtigsten Arbeitsbedingungen bei der spanabheben-den Formgebung (Additional Discussions of Determination of Optimal Machining Conditions for Metal-cutting)," *Werkstatt und Betrieb,* Vol. 90, No. 1 (1957).

Woldman, N. E., and R. C. Gibbons. *Machinability and Machining of Metals.* New York: McGraw-Hill Book Company, Inc., 1951.

Wu, S. M., and D. S. Ermer. "Maximum Profit as the Criterion in the Determination of the Optimum Cutting Conditions," *Transactions of the ASME-B,* Vol. 88, No. 4 (November, 1966).

———, and ———. "The Effect of Experimental Error on the Determination of the Optimum Metal-Cutting Conditions," *ASME Paper No. 66-WA/PROD-2* (1966).

———, and ——— and W. J. Hill. "An Exploratory Study of Taylor's Tool-Life Equations by Power Trans-formations," *Transactions of the ASME,* Vol. 88, No. 1 (February, 1966).

Zlatin, Norman, et al. *Final Report on Machinability of Metals, AFML-TR-65-444 (AD 478 214).* Cincinnati: Metcut Research Associates, Inc., 1966.

———, et al. *Final Report on Machining of Refractory Materials, ASD-TDR-581 (AD 414 988).* Cincinnati: Metcut Research Associates, Inc., 1963.

———, et al. *Machinability Parameters on New and Selective Aerospace Materials, AFML-TR-69-144 (AD 843 359).* Cincinnati: Metcut Research Associates, Inc., 1969.

———, et al. *Machining of New Materials, AFML-TR-67-339 (AD 824 483).* Cincinnati: Metcut Research Associates, Inc., 1967.

Zorev, N. N. *Metal Cutting Mechanics.* Trans. by H. S. H. Massey. New York: Pergamon Press, 1966.

———. "The Effect of Tool Wear on Tool Life and Cutting Speed," *Russian Engineering Journal,* Vol. XLV, No. 2 (1965).

INDEX

A

Abex machinability data system, 64, 114–21, 129

Acceleration cycles, 116

Accuracy, 1, 5, 130, 154, 165, 174

ADAPT, 22–23

ADAPT/AUTOSPOT, 16–17, 21–22, 38, 85, 86

Adaptive constraint control, 165

Adaptive control, 58, 160, 163–166, 174

Adjustment factors, 115, 116, 127

Air Force Machinability Data Center, 126, 130, 186

Air Force Manufacturing Technology Division, 186

AISI B1112 steel, 67, 172

AISI 4340 steel, 148–52, 178

Aluminum, 7, 31, 186

Analog control, 22

Approach angle, 108, 113

APT, 16, 22, 85, 86, 97, 113–14, 160, 162

Arc cutting, 21

Automatic tool changers, 148

Axial turning, 77, 78–79, 97 (*see also* Turning)

B

Ball end mill, 116

Base cutting-condition data file, 116

Bilogarithmic graphs, 66–67, 69, 154, 176, 177

Block, data, 17–18, 20

Boring, 20, 31, 86, 122

Boring bars, 122

Brass, 31

Broaching, 173

Bronze, 31

C

Cast iron, 31, 67, 176, 186

Cathode ray tube display, 87, 162, 163

Center drilling, 86, 89, 141, 142

Chamfering, 141, 142

Chip breakers, 104

Chip clearance, 161

Chip cross-sectional area, 103–104, 108, 109, 127, 158

Chip formation, 174, 176

Cobalt alloys, 186

Coding form, *see* Programming manuscript

Coefficient of friction, 174

Collision checks, 97, 108

Computer cards, 38, 82, 87, 130, 134, 140

Computer models, mathematical, 63, 99, 115, 139, 165, 166

Computers, 15, 39–40, 58, 64–65, 79–86, 97–99, 104, 114, 115, 118–25, 126, 127, 128, 129, 130, 134, 142–143, 144, 157–60, 161–62

Computer time, 99, 109

Constant-pressure lathes, 173, 177

Contouring, 16, 17, 20, 21, 22–23, 38, 97

Control systems, numerical, 10, 18
  sophistication of, 20–23, 58

Control tape, 5, 14, 15, 17, 18, 21, 38, 40, 41, 58, 154, 163

Control units, numerical, 161, 162, 163

Coolants, *see* Cutting fluids

"Cooling-off" period, 11

Coordinate axis locations, 17, 20, 21, 22, 108–109

Copper alloys, 31

Core-drilling, 31, 86, 80

Costs, 135, 137–39, 140, 145, 151–52, 165
  analysis of, 130, 144–152
  blade reset, 120, 142

Costs (*continued*)
  cutting, 76
  cutting-tool, 5, 6, 137, 157, 167
  development, 126
  grinding-wheel, 139, 142, 152
  handling, 144, 145, 166–67
  idle, 73–76
  insert, 139, 142, 151, 152
  labor, 105
  machine-tool, 167
  machine-tool time, 129
  machining, 115, 118, 119–21, 125, 127,
    140–42, 157, 166, 173, 186, 187
  manufacturing, 118–21
  operation, 130, 143, 145, 157, 167
  part, 1, 73–76, 105, 121, 140, 152
  programming, 126
  setup, 144, 145, 157
  tool changing, 76, 78, 151, 166–67
  tool presetting, 151, 152
  tool rebrazing, 120, 162
  tool reconditioning, 137–39, 141, 142, 151,
    152, 155
  tool regrinding, 76, 78, 120, 142
  tool replacement, 137, 151, 152
Counterboring, 31, 86
Countersinking, 86
Crater wear, 64, 104 (*see also* Tool wear)
Cut dimensions, 25 (*see also* Depth of cut,
  Length of cut, Width of cut)
CUT statements, 108
Cutter location data, 106
Cutting fluids, 25, 34, 39, 43, 65, 67, 83, 135,
  137, 143, 172, 173, 177 178, 188
Cutting forces, 25, 104, 116, 154, 160, 164,
  173–74, 177
Cutting-speed constant, 66, 76
Cutting-speed ratio, 108
Cutting time, 76, 104, 105, 119
Cutting-tool file, 87, 90, 97–99, 109–13, 159
Cutting-tool preset dimensions, 10
Cutting tools, 3, 5, 12, 18, 26, 129
  adapters for, 90
  analysis of, 93
  availability of, 10
  brazed carbide, 81, 128, 139, 142, 152, 173,
    184
  breakage of, 2, 127, 154
  carbide, 68, 69, 78, 116, 169–71, 173, 176,
    178, 180
  carbide insert, 81, 128, 139, 142, 145, 148,
    173, 184
  cemented carbide, 67, 79, 116
  cemented oxide, 67
  ceramic, 170–71
  classification of, 92
  cobalt steel, 173
  costs of, 119, 120–21, 137–39
  descriptions of, 15, 21, 23, 118
  geometry of, 10, 12–13, 23, 25, 34, 35, 43,

Cutting Tools (*continued*)
    86, 87, 90, 103, 109, 116, 118, 122, 134,
    135, 137, 143, 145, 148, 158, 160, 172,
    173, 177, 180
  grinding of, 158
  high-speed steel, 67, 78, 116, 142, 148, 152,
    169–71, 172, 173, 175, 176, 178, 180, 184
  length of, 12, 104, 108
  limits of, 109
  location of, 108
  manufacturers of, 115
  materials, 23, 34, 35, 130, 135, 137, 145,
    148, 162–63, 172, 177, 180
  milling, 87
  parameters of, 10, 25–26, 135, 137
  paths of, 43, 64, 96, 97
  presetting, 139
  purchase of, 188
  resharpening, 139
  selection of, 18, 21, 39, 63, 79–80, 83, 86,
    96, 97, 99, 159, 164
  shipments of, 169
  standard, 102–103
  variety of, 85
Cycle time, 6, 10

D

Data, definition of, 15
Data-collection systems, 26
Data files, 91, 97–99, 119, 130–31, 134 (*see
  also* Base cutting-condition data file, Cut-
  ting-tool file, Experimental data file,
  Machine-tool file, Material file, Plant cost
  data file, Shop-generated data file, Time-
  study and cost file, Tool geometry file)
Data formats, 2, 17–23, 62
Data-handling machines, 15, 17, 18
Data management, 109–13
Data sets, 143, 145, 148
Data storage, 60, 115–16
Deceleration cycles, 116
Decision tables, 115
Deep-hole drilling, 86
Deflection, 23
  of cutting tools, 7, 25, 127, 163, 164, 165
  sensors, 58
Deformation, 2
Depreciation, 78, 120
  of cutting tools, 152
Depth of cut, 21, 65, 72, 73, 83, 93, 94, 103,
  104, 108, 109, 113, 115, 116, 118, 122,
  135, 137, 139, 143, 145, 148, 160, 163
Direct numerical control, 160–63
Disk file, computer, 69, 80, 82, 87, 161, 162
Down time, 1, 20, 167
Drilling, 5, 20, 30, 86, 89, 93, 97, 99–103,
  119, 130, 148, 151, 152, 159, 160, 173,
  176, 184, 186
Ductility, 83

Dynamometers, 174

E

Electronics Industries Association, 17, 22
Empirical data, 97, 130
Emulsion coolant, 67
End milling, 26, 119, 173, 184
End-mill slotting, 130, 141, 186
End thrust, 174
Equations
  cost and time, 141
  generalized cost, 142
  optimum machining, 152–54
Ernst, Hans, 169
Europe, 96
EXAPT machinability data system, 64, 96, 105–106, 109–14, 129, 158–60
Experimental data file, 116, 121
Extra tool travel, 65, 139, 145, 148

F

Face milling, 26, 130, 141, 148, 173, 178, 186
Facilities, 187
FAST machinability data system, 63–64, 86, 93–95
FASTER machinability data system, 86, 90, 95
FASTEST machinability data system, 86, 95
Feed, 3, 5, 10, 12, 18, 20, 21, 23, 31, 34, 38, 43, 64, 65, 72, 76, 79, 80, 83, 86, 87–89, 93, 94–95, 99–102, 103, 104, 107, 108, 109, 113, 115, 116, 119, 121, 122, 127, 129, 130, 135, 137, 141, 143, 145, 148, 151, 154, 155, 159, 160, 163, 164, 172, 173, 177, 178
Feedback, 14, 164
Feed-rate/drill diameter relationship, 89, 102
Field, Michael, 176
Filing, 169
Finish machining, 11, 109, 116, 144–45, 148
Fixtures, 12, 18, 23, 39, 41, 43, 129, 161, 164
Flank wear, cutting tool, 69, 81
Flank wear land, 69, 104, 116, 121
Floor space, 5, 187
Flute length, tap, 93
Foreman, 118, 122, 125, 188
FORTRAN, 22–23, 86, 122–23
Function names, 91
Funding, 125–26

G

Gages, 18, 179
General Electric machinability data system, 63, 105, 129
Geometrical processor (EXAPT), 108
Geometry
  cutting tool, 10, 12–13, 25, 34, 35, 43, 64,

Geometry (*continued*)
  86, 90, 102–103, 109, 116, 122, 134, 135, 137, 143, 145, 148, 158, 160, 172, 173, 177, 180
  part, 23, 76, 107, 115, 122, 135, 139, 140, 155, 160
  workpiece, 64, 96–97, 160
Gilbert, W. W., 65, 105, 115
Grinding, 169

H

Handbooks, 26, 34–35, 186
Handling time, 5, 141
Hardness
  cutting-tool, 64, 116–18, 175, 177
  workpiece, 34, 58, 64, 67, 69, 79, 80, 119, 134, 140, 142–43, 144, 148, 158, 160, 163, 177, 184
Heat distortion, 11
Heat treatment, 68
High-efficiency (Hi-E) machining, 77–79, 81, 84 (*see also* Optimized machining)
High-speed steel, 67 (*see also* Cutting tools, high-speed steel)
Honing, 169
Horsepower, 25–26, 43, 65, 78–79, 80–81, 84, 89, 115, 116, 154, 155, 165, 174 (*see also* Power, Unit horsepower)
Horsepower sensors, 58

I

IBM System/360, 22
Idle time, 76
Indexing, 145
Input data, computer, 21, 22–23, 41, 87, 93, 114, 118, 140
Input data sheet, 68–69, 79–84
Input media, 5, 35 (*see also* Computer cards, Control tape, Keyboard input, Magnetic tape)
Index file cards, 99, 105, 108, 109 (*see also* Data files)
Interference, cutting-tool, 161
Internal shear, 174

J

Jigs, 18

K

Keyboard input, computer, 87, 162, 163

L

Labor rate, 76, 78, 137, 155, 158, 169
Language, machine, 15, 35, 39, 62, 90–92, 97, 160
Lapping, 169

Lead angle, 80
Lead time, 10
Length of cut, 139, 141–42, 145, 148
Linear interpolation, 22–23, 97
Loading and unloading, 140, 143, 157, 184
    time for, 119, 137
Logic, computer, 64, 83, 121–25, 126
Long-cycle operations, 119
Lot size, 10, 142–43

M

Machinability
    concepts of, 127
    definition of, 1
    parameters of, 64–65 (see also Machining conditions)
    problems of, 96–97, 115
    systems analysis of, 64
    variables of, 34, 41–43, 58 (see also Operational disturbances)
Machinability data, 115–16, 144
    accuracy of, 20, 162
    acquisition of, 134
    adequate supply of, 1
    availability of, 162–63, 169
    categories of, 2, 25–26
    contributions of, 188
    costs of, 126
    empirical, 63–65
    evaluation of, 43–58
    experience with, 26–30, 130
    generalized, 40–41, 160
    importance of, 167, 169
    for N/C, 184–86
    need for, 163–64, 186
    parameters, 2, 3 (see also Machining conditions)
    quality of, 41, 172
    significance of, 187–88
    sophisticated, 184
    sources of, 2, 5, 18–20, 23, 26–35, 60, 85–86, 118
    standards, 30, 34
    testing of, 116
    use of, 10, 15, 157, 160, 186
    variables of, 23–25
Machinability data systems
    computerized, 2, 64–64, 125, 126 (see also particular system)
    generalized, 2–3
    importance of, 1–2
    simple, 2–3, 5
    sophistication of, 2
    specialized, 3–10
    variety of, 63
Machinability index, 172–73, 178
Machinability ratings, 30–35, 69, 80, 82–83, 87 (see also Machinability index)

Machinability tests, 173–77
Machine-shop supervisors, 188
Machine-tool builders, 18, 26, 116
    specifications from, 26, 30–34
Machine-tool cutting data, 26
Machine-tool file, 87, 89, 94, 97–99, 109–13, 116, 121, 155, 159
Machine-tool operators, 9–10, 13, 58, 121, 127, 134, 172, 188
Machine tools
    acquisition of, 18, 188
    capacity of, 1
    capabilities of, 3–5, 18, 20–23, 26, 30
    characteristics of, 25, 65, 89, 115, 167
    condition of, 35
    configuration of, 6
    conventional, 3, 9, 118, 173, 184
    costs of, 1, 3, 105, 137–39
    descriptions of, 38, 116
    economics of, 10
    efficiency of, 20, 129
    limitations of, 26, 103, 109
    multiple-spindle, 9
    numbers of, 169
    operation of, 10
    producibility with, 9
    requirements, 187
    selection of, 6–7, 164
    settings, 25
    sophistication of, 20–23, 58, 184
    use of, 3–5, 161
    variety of, 5
Machining, 10, 11
    conventional, 1, 2, 23
    costs of, 186, 187
    economics of, 219, 169–72
    foundations of, 1
    interruptions in, 154
    repetitive, 23
    work done in, 174
Machining centers, 5, 7, 85, 86, 105, 148–52
Machining conditions, 10, 15, 17, 23, 65, 85–86, 93, 96, 99, 104, 106–109, 113, 114, 115, 121, 122, 127, 130, 135–37, 140, 159, 160–61, 163, 165, 172, 173–79, 180, 184
Machining Data for Numerical Control, 186
Machining Data Handbook, 34, 184
Machining "strategy," 163, 165
Machining time, 2, 3, 10, 13, 31, 114, 157–58
Machining variables, 34, 41–43, 58 (see also Operational disturbances)
Macros, 23
Magnetic disks, see Disk file, computer
Magnetic tape, 87 (see also Input media)
Maintenance subroutines, 114 (see also Service programs)
Major vocabulary words, 90–91, 92
Management, 161, 162, 169, 171–72, 186–88

Manufacturing
  conventional, 184
  objectives of, 2
  philosophy of, 7, 8, 11, 23
  small-lot, 167
Manufacturing engineers, 10, 118, 122, 125,
  128, 186, 188
Manufacturing records, 60
Manufacturing time, 86
Marketing, 170
Mass production, 7–8
Material file, 68, 87–89, 94, 97–99, 103, 109,
  113, 155
Materials, 5, 6, 10, 105, 116, 140, 154–55 (see
  also particular material)
  cutting-tool, 34, 35, 43, 64, 68–69, 103,
    116, 118, 135, 137, 143, 145, 148,
    162–63, 172, 177, 180
  deformation of, 2
  limits of, 109
  shear of, 2
  surface condition of, 67–68
  variety of, 184
  workpiece, 30–35, 58, 64, 80, 87–89, 103,
    115, 116, 119, 134, 142–43, 163, 172,
    177, 188
Maximum production, 5, 7, 8, 10–11, 63, 64,
  73, 77, 84, 115, 129, 130, 144, 151,
  152–54, 186
  speed for, 65, 77
  tool life for, 77, 105
Metal-removal rates, 25, 26
Metcut Research Associates, Inc., 115
Microstructure, 34, 64, 67, 177
"Milepost" project documentation, 125
Milling, 20, 25–26, 31, 77, 86, 87, 93, 94–95,
  97, 119, 135, 139, 141, 142, 152
Minimum part cost, 1–2, 5, 8, 10–11, 63, 64,
  73, 76–77, 84, 115, 121, 127, 129, 130,
  144, 145, 151, 152–54, 186
  speed for, 65, 76–77
  tool life for, 77–78, 105
Minor vocabulary words, 90, 91
Miscellaneous functions, 17, 21
Modified electric typewriter, 15
Modulus of elasticity, 7
Molybdenum steel, 31
Motion control, 23
Motion errors, 43

N

Nickel alloys, 186
Nomographs, 25–26
Nonmetallic materials, 186
Normal force, 174
Nose radius, 72, 73, 80, 108, 118

O

Oil coolant, 67
Operational disturbances, 23–25, 41–43
"Operational intensity," 105
"Operation family" concept, 115, 118, 119
Operations sequence, 39, 64, 96, 97
Operation times, 64, 86, 130, 137, 141, 145,
  148, 152, 167
Operation type, 25–26
Optimal adaptive control, 166
Optimized machining, 10–11, 14, 63, 73–77,
  85, 97, 105, 118, 121, 127, 129, 130,
  152–54, 155, 160, 165, 167, 184–86 (see
  also Maximum production, Minimum
  part cost)
Order quantity, 10
Overhead rate, 76, 78, 119, 121, 127, 137,
  155, 158, 169

P

Pallet shuttling, 21, 22, 95
Parabolic interpolation, 22–23
Part analysis, 125
Part design, 6–7, 10, 11
Part drawings, 5, 18, 39, 160
Parts processing, 2, 10, 11, 18
Parts programmers, 10, 13, 18–19, 20, 25,
  35–40, 41, 43–58, 60, 62, 86, 93, 97, 108,
  115, 125, 126–127, 129, 159, 164, 186,
  188
Parts programming, 3, 7–8, 10, 18, 105–108,
  109, 161
  computer-aided, 10, 38, 41, 160
  cost of, 5, 113–14
  manual, 15, 39, 41, 58, 114, 160
Parts programs, see Programming systems
Peck drilling, 103
Penetration rate measurement, 177
Performance index, 165, 166
Peripheral end milling, 130, 141, 151, 186
Plant cost data file, 116, 118
Positioning, 17, 20–21, 97
Postprocessor, 38, 106, 107
Power, 1, 103, 107, 160 (see also Horsepower,
  Unit horsepower)
Premachining operations, 97
Preparatory functions, 17, 21
Preset dimensions, cutting tool, 10
Process engineers, 10, 39, 122, 125, 127, 129,
  188
Process sequence, 18, 21, 114, 155
Producibility, 6, 9
Production analysis, 144–52
Production planners, 188
Production rate, 140, 141, 148, 152, 165, 187
Production quantity, 5, 10, 140

Production schedules, 18
Production time, 2, 20
Profile factor, 73, 80, 83
Profiling, 21
Programmed adaptive control, 165
Programming manuscript, 17, 18, 22, 35–38, 39
Programming systems, 15–17, 85–86, 108, 109, 114, 123–25, 155
  general, 15–17
  requirements of, 15
  sophistication of, 39–40, 58
  special-purpose, 15–17
Project planning, 125
Prototype production, 7–8
Purchasing agents, 188

Q

Quality, part, 7

R

Rapid traverse, 105–106, 116, 119, 139, 141, 145, 148
Reaming, 31, 89, 93, 130, 139, 141, 142, 152, 186
Refractory alloys, 186
Reporting systems, 161
Research, 26
Rigidity, 121, 158, 174
  of cutting tools, 5, 6, 7, 12
  of fixtures, 6, 7, 12
  of machine tools, 65
  of parts, 6, 7
Roughing, 11, 107–108, 109, 116, 144, 145
Routing sheet, 18

S

Salesmen, 172
Sand castings, 68
Semifinishing, 11, 144, 145
Sensitivity analyses, 121, 127–28
Sensors, 58, 163
Service programs, 109–13 (see also Maintenance subroutines)
Setup, 7, 11, 18, 21, 41, 97, 119, 137, 140, 141, 143, 157, 158
Shear, 2, 174
Shear angle, 174
Shop-generated data file, 116, 118, 119
Short-cycle operations, 115
Side cutting-edge angle, 13, 73
Side milling, 31
Simple feedback control, 164
Slide velocity, 164
Slope cutting, 21
Software, see Programming systems
Speed/tool-life relationship, 43, 66–67, 69–72,

Speed/tool-life relationship (continued)
  79, 80–81, 104, 109, 135, 137, 140, 145, 148, 154, 158, 186
Speeds
  cutting, 1, 3, 10, 12, 18, 20, 21, 25–26, 30, 34, 38, 43, 64, 66, 67, 69, 76, 77, 83, 86, 99–102, 104, 108, 109, 113, 115, 116, 122, 129, 130, 143, 144, 145, 148–51, 152, 155, 158, 159, 160, 164, 167, 172, 177, 178
  machine-tool, 5, 26, 58, 65, 79, 80–81, 84, 87, 89, 93, 94, 95, 105–106, 108, 116, 121, 159, 163
Speeds and Feeds for Better Turning Results, 34
SPLIT, 15–16
Stainless steel, 13, 184, 186
Starting data, 180–84, 186
Steel, 7, 31, 67, 176, 184, 186
Strain gages, 174
Subroutines, computer, 39, 87, 118, 119
Surface finish, 1, 5, 13, 20, 25, 26, 43, 64, 68, 72, 80, 83, 109, 115, 116, 118, 122, 130, 154, 160, 165
Surface integrity, 13, 67–68, 130, 154
Symbolic program system (SPS), 21–22

T

Table indexing, 20, 21, 22
Tab sequential format, 18, 21
Tangential load, 25–26
Tapping, 13, 20, 31, 86, 93, 130, 135, 139, 141, 142, 144, 145, 152, 173, 176, 179, 186
Taylor, F. W., 66, 105
Taylor's equation, 66, 67, 104, 109, 115, 152, 154, 155, 158, 176, 177
Technological processor (EXAPT), 97, 99, 103, 108, 109, 113
Temperature, cutting, 163, 170, 175, 176
Temperature sensors, 58
Tensile properties, 34
Testing, 13, 116
  of control tapes, 43–58
  of machinability data, 65, 116
Thermal variations, 23
Thermocouples, 173, 175–76
The Tenth American Machinist Inventory of Metalworking Equipment, 169
Time-study and cost file, 155
Time-study data, 130, 135, 137–39, 140, 145–48, 151–52
Time-study engineers, 188
Titanium alloy, 144, 179
Tolerance, 10, 23, 64, 81, 115, 116, 121, 122, 163
Tool analysts, 10, 13
Tool changing time, 20, 21, 22, 95, 105, 116,

Tool changing time (*continued*)
120, 148, 157, 167
Tool designers, 188
Tool dynamometers, 173–74
Tool engineering, 38–39, 188
Tool geometry file, 87, 90
Toolholders, 3, 10, 26, 90, 118, 174
Tool indexing, 41, 137
Tool life, 1, 2, 6, 7, 10, 13, 20, 25, 26, 43, 58, 65, 66, 77–78, 83, 105, 109, 115, 116, 130, 137, 142, 143, 144, 145, 148, 152, 154, 158, 165, 167, 176, 178, 179–80, 184, 187–88
Tool-life end point, 135, 137, 179–80 (*see also* Flank wear land, Tool wear)
Tool-life tests, 130, 173, 177–84, 187, 188
accelerated, 176–77
Tool movements, 135, 139
Tool path calculation, 108
Tool steel, 186
Tool wear, 23, 64, 104, 109, 127, 163, 174
Tool-wear sensors, 58
Torque, 25–26, 103, 107, 154, 159, 160, 163, 164, 165
Total automatic control concept, 41, 58
Toughness, 170
Training
managerial, 125
programmer, 126–27
Transducers, 174
Turning, 77, 103–104, 106–109, 115, 116, 119, 130, 134, 139, 141, 142, 144–48, 152, 154, 160, 173, 180, 184, 186 (*see also* Axial turning)

Turret block, 145

U

Ultrahigh-strength steel, 186
United States Army Rock Island Arsenal Laboratory, 184
United States Army Weapons Command Metal Cutting Advisory Committee, 184
Unit horsepower (UHP), 79, 83, 116, 155 (*see also* Horsepower, Power)

V

Value analysis engineers, 188
Variable block format, 18
Vibration, 163

W

Witthof, J., 105
Width of cut, 93, 135, 137, 139, 143, 148, 163
Word, data, 17–18
Word-address format, 18, 21
Work content, 3–5, 6, 15, 20
Workpieces
configurations of, 17, 21
geometry of, 115, 122, 139, 160
machinability of, 177
materials, 15, 18, 23, 25, 34, 43, 163
requirements of, 15
specifications of, 15, 18
variations in, 41
WORK statements, 108